Review of International Technologies for Destruction of Recovered Chemical Warfare Materiel

Committee on Review and Evaluation of International Technologies for the Destruction of Non-Stockpile Chemical Materiel

Board on Army Science and Technology

Division on Engineering and Physical Sciences

NATIONAL RESEARCH COUNCIL
OF THE NATIONAL ACADEMIES

THE NATIONAL ACADEMIES PRESS
Washington, D.C.
www.nap.edu

THE NATIONAL ACADEMIES PRESS **500 FIFTH STREET, N.W.** **Washington, DC 20001**

NOTICE: The project that is the subject of this report was approved by the Governing Board of the National Research Council, whose members are drawn from the councils of the National Academy of Sciences, the National Academy of Engineering, and the Institute of Medicine. The members of the committee responsible for the report were chosen for their special competences and with regard for appropriate balance.

This study was supported by Contract No. W911NF-05-C-0078 between the National Academy of Sciences and the Department of Defense. Any opinions, findings, conclusions, or recommendations expressed in this publication are those of the author(s) and do not necessarily reflect the views of the organizations or agencies that provided support for the project.

International Standard Book Number-10: 0-309-10203-0
International Standard Book Number-13: 978-0-309-10203-2

Cover: Images courtesy of the public affairs office of the Non-Stockpile Chemical Materiel Project, U.S. Army, Chemical Materials Agency. The munitions shown illustrate the condition in which such items are often found when they are recovered from munitions burial sites.

Limited copies of this report are available from:

Board on Army Science and Technology
National Research Council
500 Fifth Street, N.W., Room 940
Washington, DC 20001
(202) 334-3118

Additional copies are available from:

The National Academies Press
500 Fifth Street, N.W.
Lockbox 285
Washington, DC 20055
(800) 624-6242 or (202) 334-3313
(in the Washington metropolitan area)
Internet, http://www.nap.edu

Printed in the United States of America

THE NATIONAL ACADEMIES

Advisers to the Nation on Science, Engineering, and Medicine

The **National Academy of Sciences** is a private, nonprofit, self-perpetuating society of distinguished scholars engaged in scientific and engineering research, dedicated to the furtherance of science and technology and to their use for the general welfare. Upon the authority of the charter granted to it by the Congress in 1863, the Academy has a mandate that requires it to advise the federal government on scientific and technical matters. Dr. Ralph J. Cicerone is president of the National Academy of Sciences.

The **National Academy of Engineering** was established in 1964, under the charter of the National Academy of Sciences, as a parallel organization of outstanding engineers. It is autonomous in its administration and in the selection of its members, sharing with the National Academy of Sciences the responsibility for advising the federal government. The National Academy of Engineering also sponsors engineering programs aimed at meeting national needs, encourages education and research, and recognizes the superior achievements of engineers. Dr. Wm. A. Wulf is president of the National Academy of Engineering.

The **Institute of Medicine** was established in 1970 by the National Academy of Sciences to secure the services of eminent members of appropriate professions in the examination of policy matters pertaining to the health of the public. The Institute acts under the responsibility given to the National Academy of Sciences by its congressional charter to be an adviser to the federal government and, upon its own initiative, to identify issues of medical care, research, and education. Dr. Harvey V. Fineberg is president of the Institute of Medicine.

The **National Research Council** was organized by the National Academy of Sciences in 1916 to associate the broad community of science and technology with the Academy's purposes of furthering knowledge and advising the federal government. Functioning in accordance with general policies determined by the Academy, the Council has become the principal operating agency of both the National Academy of Sciences and the National Academy of Engineering in providing services to the government, the public, and the scientific and engineering communities. The Council is administered jointly by both Academies and the Institute of Medicine. Dr. Ralph J. Cicerone and Dr. Wm. A. Wulf are chair and vice chair, respectively, of the National Research Council.

www.national-academies.org

COMMITTEE ON REVIEW AND EVALUATION OF INTERNATIONAL TECHNOLOGIES FOR THE DESTRUCTION OF NON-STOCKPILE CHEMICAL MATERIEL

RICHARD J. AYEN, *Chair*, Waste Management, Inc. (retired), Jamestown, Rhode Island
ROBIN L. AUTENRIETH, Texas A&M University, College Station
ADRIENNE T. COOPER, Temple University, Philadelphia, Pennsylvania
MARTIN GOLLIN, St. Davids, Pennsylvania
GARY S. GROENEWOLD, Idaho National Laboratory, Idaho Falls
PAUL F. KAVANAUGH, BG, U.S. Army (retired), Fairfax, Virginia
TODD A. KIMMELL, Argonne National Laboratory, Washington, D.C.
LOREN D. KOLLER, Oregon State University (retired), Corvallis
DOUGLAS M. MEDVILLE, MITRE Corporation (retired), Reston, Virginia
GEORGE W. PARSHALL, E.I. DuPont de Nemours & Company (retired), Wilmington, Delaware
JAMES P. PASTORICK, Geophex UXO, Ltd., Alexandria, Virginia
LEONARD M. SIEGEL, Center for Public Environmental Oversight, Mountain View, California
WILLIAM J. WALSH, Pepper Hamilton LLP, Washington, D.C.

Staff

HARRISON T. PANNELLA, Study Director
JAMES C. MYSKA, Senior Research Associate
ALEXANDER R. REPACE, Senior Program Assistant (from March 2006)
LaTANYA CLEMENCIA, Senior Program Assistant (until March 2006)

BOARD ON ARMY SCIENCE AND TECHNOLOGY

Preface

The Committee on Review and Evaluation of International Technologies for the Destruction of Non-Stockpile Chemical Materiel was appointed by the National Research Council (NRC) in response to a request by the U.S. Army's Project Manager for Non-Stockpile Chemical Materiel.

The committee's focus was on destruction technologies for recovered chemical weapons that are not now a part of the repertoire of the Project Manager for Non-Stockpile Chemical Materiel but that could prove to be useful additions or replacements. To that end, countries using or considering the use of technologies for the destruction of old and abandoned chemical weapons to meet requirements of the international Chemical Weapons Convention (CWC) treaty, along with the developers of such technologies, were contacted. This report summarizes the acquired information, evaluates the technologies to the extent possible, and presents the results. Consideration was given to technologies that might offer advantages over those now in use by the U.S. Army or those that might otherwise prove useful, especially for situations not now adequately covered, such as destruction operations where large numbers of recovered munitions must be treated. A limited effort was expended on the assessment and storage of recovered chemical weapons.

Several individuals met with visiting committee members in Europe and provided helpful information on the status of international technologies in other countries. The committee offers its thanks for their assistance:

- Richard Soilleux, Technical Leader, Defence Science and Technology Laboratory, U.K. Ministry of Defence, Porton Down, England;
- Hans-Joachim Grimsel, Managing Director, Gesellschaft zur Entsorgung von chemischen Kampfstoffe und Rüstungs-Altlasten (GEKA), Munster, Germany;
- Ralf Trapp, Senior Planning Officer, Office of the Deputy Director-General, Organisation for the Prohibition of Chemical Weapons, The Hague, The Netherlands;
- Jerzy Mazur, Head, Chemical Demilitarisation Branch (CDB), Organisation for the Prohibition of Chemical Weapons, The Hague, Netherlands;
- Jeff Osborne, Senior Substantive Officer, CDB, Organisation for the Prohibition of Chemical Weapons, The Hague, Netherlands;
- Herbert De Bischopp, Professor, Royal Military Academy, Brussels, Belgium; and
- Michel Lefebvre, Professor, Royal Military Academy, Brussels, Belgium.

The committee would also like to thank vendor representatives and others who assisted in information gathering for this report. See Appendix D for the names of these individuals.

The study was conducted under the auspices of the NRC's Board on Army Science and Technology (BAST). The BAST was established in 1982 as a unit of the National Research Council at the request of the U.S. Army. The BAST brings to bear broad military, industrial, and academic scientific, engineering, and management expertise on Army technical challenges and other issues of importance to senior Army leaders. The board discusses potential studies of interest; develops and frames study tasks; ensures proper project planning; suggests potential committee members and reviewers for reports produced by fully independent ad hoc study committees; and convenes meetings to examine strategic issues. The board members listed on p. vi were not asked to endorse the committee's conclusions or recommendations, nor did they review the final draft of this report before its release. However, board members with appropriate expertise may be nominated to serve as formal members of study committees, or as report reviewers.

The chair acknowledges the superb support of the BAST director, Bruce A. Braun, and the study director, Harrison T. Pannella. Valuable assistance was provided by James C. Myska, Alexander R. Repace, and LaTanya Clemencia of the NRC staff. In view of the international nature of the necessary information gathering, committee members were faced with considerably more challenges than is typical for a National Research Council study in the area of chemical demilitarization, and the chair is grateful for their hard work and diligence in carrying out this study.

Richard J. Ayen, *Chair*
Committee on Review and
Evaluation of International
Technologies for the Destruction
of Non-Stockpile Chemical
Materiel

Acknowledgment of Reviewers

This report has been reviewed in draft form by individuals chosen for their diverse perspectives and technical expertise, in accordance with procedures approved by the National Research Council's Report Review Committee. The purpose of this independent review is to provide candid and critical comments that will assist the institution in making its published report as sound as possible and to ensure that the report meets institutional standards for objectivity, evidence, and responsiveness to the study charge. The review comments and draft manuscript remain confidential to protect the integrity of the deliberative process. We wish to thank the following individuals for their review of this report:

William B. Bacon, Shaw Environmental & Infrastructure,
Ruth M. Doherty, Naval Surface Warfare Center,
Gene Dyer, consultant,
Jeff Edson, Colorado Department of Public Health and Environment,
Mario H. Fontana, University of Tennessee (Knoxville),
Dan Luss, University of Houston,
James F. Mathis, Exxon Corporation (retired),
Hyla S. Napadensky, Napadensky Energetics Inc.,
William R. Rhyne, ABS Consulting, Inc. (retired), and
William Tumas, Los Alamos National Laboratory.

Although the reviewers listed above have provided many constructive comments and suggestions, they were not asked to endorse the conclusions or recommendations, nor did they see the final draft of the report before its release. The review of this report was overseen by Richard A. Conway, Union Carbide Corporation (retired). Appointed by the National Research Council, he was responsible for making certain that an independent examination of this report was carried out in accordance with institutional procedures and that all review comments were carefully considered. Responsibility for the final content of this report rests entirely with the authoring committee and the institution.

Contents

Tables and Figures

TABLES

FIGURES

Acronyms

ACWA Assembled Chemical Weapons Assessment/Alternatives
ADP acid digestion process
AEL airborne exposure limit
ASME American Society of Mechanical Engineers

CAA Clean Air Act
CAIS chemical agent identification set(s)
CAMDS Chemical Agent Munitions Disposal System
CATOX catalytic oxidation (unit)
CDC controlled detonation chamber
CEB Centre d'Etudes du Bouchet
CERCLA Comprehensive Environmental Response, Compensation, and Liability Act
CFR Code of Federal Regulations
CG phosgene
CK cyanogen chloride
CO carbon monoxide
CS orthochlorobenzylidene malononitrile (tear gas)
CWC Chemical Weapons Convention
CWM chemical warfare materiel

DA diphenylchloroarsine (Clark I)
DAVINCH detonation of ammunition in vacuum integrated chamber
DC diphenylcyanoarsine (Clark II)
DCD Deseret Chemical Depot
DE destruction efficiency
DF a binary precursor (methylphosphonic difluoride)
DFS deactivation furnace system
DM adamsite
DOD Department of Defense
DOE Department of Energy
DOT Department of Transportation
DPG Dugway Proving Ground
DRE destruction and removal efficiency
DRES Defence Research Establishment Suffield
DSTL Defence Science and Technology Laboratory
DUN dunnage furnace

EDS explosive destruction system
EPA Environmental Protection Agency
ESTCP Environmental Security Technology Certification Program

GA a nerve agent (tabun)
GB a nerve agent (sarin)
GD a nerve agent (soman)
GEKA German testing facility, Gesellschaft zur Entsorgung von chemischen Kampfstoffe und Rüstungs-Altlasten
GPL general population limit

H sulfur mustard
HCl hydrogen chloride
HD sulfur mustard (distilled)
HEPA high efficiency particulate air
HMX an explosive
HN nitrogen mustard
HS sulfur mustard
HT sulfur mustard, T-mustard combination
HVAC heating, ventilation, and air conditioning

ICV In-Container Vitrification™
IDLH immediately dangerous to life and health
IUPAC International Union of Pure and Applied Chemistry

L lewisite or liter
LIC liquid incinerator
LITANS large item transportable access and neutralization system

MEC	munitions and explosives of concern
mg	milligram
MPF	metal parts furnace
NaOH	sodium hydroxide
NEPA	National Environmental Policy Act
nm	nanometer
NRC	National Research Council
NSCMP	Non-Stockpile Chemical Materiel Project
NSCWM	Non-Stockpile Chemical Warfare Materiel
OPCW	Organization for the Prohibition of Chemical Weapons
PBA	Pine Bluff Arsenal
PCB	polychlorinated biphenyl
PD	phenyldichloroarsine
PINS	portable isotopic neutron spectroscopy
PMNSCM	Project Manager for Non-Stockpile Chemical Materiel
PPE	personal protective equipment
RAP	regulatory approval and permitting
RCRA	Resource Conservation and Recovery Act
RCWM	recovered chemical warfare materiel
RDX	an explosive
RMA	Rocky Mountain Arsenal
ROD	record of decision
RRS	Rapid Response System
SCANS	single CAIS accessing and neutralization system
SERDP	Strategic Environmental Research and Development Program
SIPRI	Stockholm International Peace Research Institute
SNPE	Société Nationale des Poudres et Explosifs
SOT	statement of task
STEL	short-term exposure limit
TNT	an explosive
TPA	triphenylarsine
TSDF	treatment, storage, and disposal facility
UV	ultraviolet
VR	Russian version of VX
VX	a nerve agent
3X	level of agent decontamination (suitable for transport for further processing)
5X	level of agent decontamination (suitable for commercial release)

Executive Summary

The purpose of this study was to identify and evaluate technologies developed or refined outside the United States that could be useful in future non-stockpile chemical warfare materiel recovery and destruction operations conducted by the U.S. Army. Candidate technologies could offer complementary capabilities or even replace current equipment or approaches. The statement of task for this study charged the Committee on Review and Evaluation of International Technologies for the Destruction of Non-Stockpile Chemical Materiel with evaluating international systems, facilities, and disposal technologies currently employed or under development in countries that need them for the treatment and destruction of inventories of non-stockpile materiel. The committee was to compare those international technologies with the technologies used in the current U.S. non-stockpile chemical weapon recovery and destruction program (which are described in Chapter 1). In early committee meetings, the U.S. Army's non-stockpile staff also asked the committee to report on any promising international technologies for assessment of chemical weapon burial sites and the assessment of recovered chemical munitions.

The United States is a signatory to the Chemical Weapons Convention (CWC), which prohibits the use of chemical weapons and mandates the elimination of existing declared stockpiles by April 29, 2007, with the possibility of a 5-year extension. This mandate applies to chemical warfare materiel (CWM) that has been recovered from sites where it had in the past been buried. In the United States, such material is referred to as non-stockpile chemical warfare materiel (NSCWM). The CWC requires the declaration and destruction of such materiel within the CWC treaty deadline if it is unearthed prior to the deadline. The CWC allows signatory nations to exclude this CWM as long as the materiel remains buried. However, when this CWM is unearthed, it becomes recovered CWM, or RCWM, and must be destroyed. The CWC allows some negotiation of the timetable for the disposal of declared CWM, although generally it should be "destroyed as soon as possible."

As of 1996, the U.S. Army had located 168 potential CWM burial sites at 63 locations in 31 states, the U.S. Virgin Islands, and the District of Columbia. The universe of buried non-stockpile CWM includes several sites where large amounts of buried CWM are located—Redstone Arsenal, Alabama; Rocky Mountain Arsenal, Colorado; Aberdeen Proving Ground, Maryland; and Deseret Chemical Depot, Utah. Medium to large amounts of buried CWM may exist at several other sites.

Obsolete chemical weapons that have been in storage since the decades following World War II constitute the U.S. chemical stockpile and are differentiated from non-stockpile materiel. Facilities in the United States that have been constructed to destroy this stockpile employ assembly line systems for separating the agent from the munition. This is feasible because the munitions are overwhelmingly in a good and consistent condition. Leakers and other occasional nonuniform munitions that are periodically encountered can cause problems out of proportion to their numbers, however.

Non-stockpile munitions, by contrast, are more typically characterized by their poor condition from having been buried for decades. As in the United States, munitions recovered from burial sites (and battlefields) in Germany, Belgium, Italy, and France exhibit a lack of uniformity regarding geometry, agent type, fired, fuzed, empty, full, corroded, and country of origin. A major focus of this study was to learn how these countries are now dealing with the recovery and destruction of these munitions and what, if any, new technologies they are considering implementing in the future. In these countries, no assembly line system exists for disassembling recovered munitions to separate the explosive from the agent. Any disassembly that has taken place has utilized various approaches, including manual positioning in machines, automatic cutting, and manual emptying of agent.

The committee considered two approaches for removing munitions from large burial sites. It concluded (see Chapter 2) that a remove-and-dispose approach is to be preferred to a remove-store (in an intermediate holding facility)-dispose approach. The remove-and-dispose approach would minimize handling and storage of potentially deteriorated munitions, thus lowering risks.

Current technologies used by the U.S. Army's Non-Stockpile Chemical Materiel Project (NSCMP) will also be applicable to the destruction of munitions recovered in the future. However, these technologies are limited in terms of the size of munition they can handle and their processing rate. The NSCMP's explosive destruction system (EDS) is a well-proven system, but individual units can only deal with relatively small munitions at a slow rate. Other technologies are suited only to deal with small quantities of agent, e.g., chemical agent identification sets (CAIS). Therefore, one goal of this study was to identify international technologies that would destroy recovered munitions at a faster rate than existing NSCMP technologies in the event that the Department of Defense (DOD) decides, as a matter of policy or as required by law, to remove large numbers of buried CWM within a relatively short period of time. In selecting these technologies, DOD would benefit from consultation with regulators and public stakeholders, particularly because of the close relationship between the choice of technology and the rate at which buried CWM can be recovered and destroyed.

EVALUATION CRITERIA

The committee attempted to focus its evaluation activities on the international chemical materiel destruction technologies that appeared to be most promising. This selection was accomplished using a tiered matrix (described in Chapter 1). The more promising technologies were placed in Tier 1 and were evaluated in detail, whereas other technologies were placed in Tier 2 and received either a lesser or only a cursory evaluation. The committee concentrated its efforts on destruction technologies suited to anticipated situations for non-stockpile CWM that has yet to be recovered. In particular, the committee was interested in examining technologies that could be implemented at sites where large quantities of buried materiel can be expected and where, consequently, higher throughputs might be desired than are achievable with current NSCMP equipment. The committee further divided the technologies into (1) those that could treat an entire munition and (2) those that destroy agent only. In evaluating the Tier 1 technologies, the following evaluation factors were employed:

- *Process maturity.* This factor is used to assess whether a particular technology has been sufficiently developed and has accumulated enough operational experience so that it can be reasonably claimed that all significant issues are understood and operation of the technology is routine.
- *Process efficacy/throughput.* This factor is used to assess whether a particular technology is fully effective in achieving its task and how efficient it is in destroying munitions or agent in terms of processing rate and/or the maximum size of munition that can be handled.
- *Process safety.* This factor is used to assess whether the technology is safe to operate, presuming that the design criteria are not exceeded and the defined operating procedures are followed.
- *Public and regulatory acceptability in a U.S. context.* This factor is used to assess whether, even though the technology may be in use in another country, it is likely to be acceptable to local community stakeholders in this country and jurisdictional regulatory bodies with specific environmental and political concerns.
- *Secondary waste issues.* This factor is used to assess whether any secondary waste streams generated by the technology present a particular problem in terms of disposal and treatment.

Costs associated with purchasing and operating a given technology would also be a significant criterion, but the committee did not have access to capital or operating cost data.

TIER 1 INTERNATIONAL TECHNOLOGIES FOR MUNITION PROCESSING

The three international technologies assigned to Tier 1 are described and discussed in Chapter 4. They do not disassemble the munition and separate the agent and the explosive but rely instead on destroying the munition and its contents in their entirety and without disassembly. They do this in one of two distinct ways:

- Cold detonation, in which an explosive donor charge is placed around the munition. The munition(s) is placed within an explosive containment structure and the donor charge detonated. The resulting pressure, temperature, and fireball destroy the explosive and agent. Offgases pass to a treatment system. In the technology summaries that follow, the controlled detonation chamber (CDC) and DAVINCH (detonation of ammunition in a vacuum integrated chamber) work this way.
- Hot detonation, in which the munition is inserted into a hot kiln (externally heated). The temperature in the kiln results in a deflagration, detonation, or burning of the munition's explosive fill, again followed by agent destruction. Offgases pass to a treatment system. In the technology summaries that follow, the Dynasafe static kiln works this way.

Table ES-1 provides summary ratings of these Tier 1 international munitions processing technologies for the five evaluation factors noted above as well as comparative rat-

TABLE ES-1 Evaluation Factor Rating Comparison of Tier 1 Munitions Processing Technologies with U.S. EDS

	Evaluation Factors (Rating[a])[b]				
Technology	Process Maturity	Process Efficacy/Throughput	Process Safety	Public and Regulatory Acceptability in a U.S. Context	Secondary Waste Issues
U.S. EDS	+	+	+	+	0
CDC	+	+	+	0	0
DAVINCH	+	+	+	0[c]	+
Dynasafe	+	+[d]	+	0	0

[a]Legend: +, acceptable; 0, partially acceptable; –, unacceptable; ?, inadequate information.

[b]Costs associated with purchasing and operating a given technology would also be a significant criterion, but the committee did not have access to capital or operating cost data.

[c]DAVINCH is more likely to be acceptable to the public than the CDC and Dynasafe because of its demonstrated ability to hold and test waste gases, but it has not yet been permitted (see the section "Public and Regulatory Acceptability in a U.S. Context" in Chapter 4).

[d]Rating is contingent on the ability of the Dynasafe process control system to confirm agent destruction in all munitions that do contain agent.

ings for the U.S. Army's EDS. Please refer to Chapter 3 for a full explanation of the criteria and ranking symbols used by the committee. Refer to the text of Chapter 4 (and to Appendix B) to learn what kind of information formed the basis for a particular ranking. Table ES-2 briefly provides engineering parameters that contributed to the rankings for the detonation technologies and the NSCMP EDS technology that are given in Table ES-1.

Controlled Detonation Chamber Technology

The CDC, an earlier version of which was originally developed in the United States, was subsequently refined in Europe and is being used there, particularly in Belgium and, to a lesser extent, in the United Kingdom. It has three main components: a blast chamber, an expansion chamber, and an emissions control unit. The blast chamber, in which the detonation occurs, is connected to a larger expansion chamber. A projectile wrapped in explosive is mounted in the blast chamber. The floor of the chamber is covered with pea gravel, which absorbs some of the blast energy. Bags containing water are suspended near the projectile to help absorb blast energy and to produce steam, which reacts with agent vapors. After the explosive is detonated in the blast chamber, the gases are vented to the emissions control system. Systems with capacities ranging from 12 pounds of TNT-equivalent (the T-10 model) to 60 pounds of TNT-equivalent (TC-60 model) have been constructed and operated. The latest versions incorporate a mechanical system to move explosive-encased munitions from the preparation area into the blast chamber. The offgas treatment system includes a reactive-bed filter to remove acidic gases and a porous ceramic filter to collect particulates like soot and dust from the pea gravel. A catalytic oxidation (CATOX) unit oxidizes CO and organic vapors from the gas stream before it is vented through a carbon adsorption bed.

The CDC appears to be well suited for destroying a range of either chemical or conventional munitions. It has been used in a production mode by the Belgian military to destroy RCWM at its test facility at Poelkapelle. At the time this report was being prepared, development work on the CDC was continuing to demonstrate the usefulness of the CDC for recovered chemical operations in the United States. The destruction efficiency of the post-detonation environment in the blast and expansion chambers appears to be over 99 percent. No published overall destruction and removal efficiency (DRE) figure has been found, but available information indicates that the CDC is capable of achieving DREs of greater than 99.9999 percent, a satisfactorily high number in the opinion of the committee. The CDC does not, however, qualify as a hold-and-test system like the EDS (described in Chapter 1) because the CDC is a flow-through system and offgases are not held and analyzed before release.

Because there is no time-consuming neutralization step as in the EDS, the CDC's throughput could be much higher than that of the EDS, which conducts only one detonation every other day. The EDS-1 can handle three mortar rounds per shot, and the EDS-2 has destroyed as many as six rounds per shot. The CDC has demonstrated destruction of two munitions per shot and could potentially destroy 40 projectiles per 10-hour shift. The current CDC also has the advantage of generating little or no liquid waste that requires subsequent processing, in contrast with the significant neutralent and rinsate effluents produced by the EDS. Pea gravel is a secondary waste that must be disposed of.

Manual operations are now minimized by slipping precast donor explosives over the projectile and mechanically moving the round into the detonation chamber. The substitution of hot air purging for washing the chamber and detonation debris with decontamination solution eliminated a set of operations that probably posed a significant risk of exposure to chemical agent.

TABLE ES-2 Specific Engineering Parameters for Existing Munitions Processing Technologies

Technology Model	Throughput Rate	Destruction Verification Capability	Largest Munition	Reliability/Operability	Transportability
EDS-2	1 detonation every other day; up to 6 munitions per detonation	Liquid and gaseous effluents can be held and tested before release	5 lb TNT-equivalent; wide range of weapons acceptance; maximum: 155-mm projectile; physical size of munition determines throughput rate	Extensive experience with chemical munitions	Fully transportable; 1 trailer
CDC (TC-60)	Up to 20 detonations per 10-hr shift; estimated potential throughput given by technology proponent as 22-40/day; actual will be determined in 2006	Monitoring of offgas prior to release to carbon filter system	60 lb TNT-equivalent; 210-mm projectile	Extensive experience with conventional munitions; has demonstrated reliability; 4 years experience in production mode without failure	Transportable on 8 tractor trailers
DAVINCH (DV-60)	Yellow bombs: 9/day Red bombs: 18/day 75-mm, 90-mm munitions: 36/day	Detonation gases held in tank and tested for agent before decision made to release or provide additional treatment	65 kg TNT-equivalent; expected to be an 8-in. projectile or a small bomb	Experience with destruction of 600 Japanese Red and Yellow chemical bombs containing various agents	DV-60 designed to be a fixed facility, not transportable
Dynasafe (SK2000)	Varies greatly with munition and operating mode; if used as an open system (continuous mode), sample throughput rates are 20/day for 8-in. projectile, 40/day for 155-mm projectile, 120/day for 105-mm projectile and 4.2-in. mortar round	Open system (continuous mode): none prior to offgas treatment; closed system (batch mode): hold and test in expansion tank	5 lb TNT-equivalent; 8-in. projectile, if fragment shield used to protect chamber; up to 750-lb bomb if most of agent is drained first	Extensive experience with conventional munitions; some experience with German chemical munitions	SK2000 designed to be a fixed facility, not transportable

DAVINCH Technology

The DAVINCH technology, developed by Kobe Steel in Japan, uses a large detonation chamber in which chemical munitions and their contents are destroyed when donor charges surrounding the munitions are detonated under a near vacuum. Although the process does not require use of a reagent to destroy the agent—accomplished by a shock wave, expansion and thermal heating from the detonation gases, and a fireball in the chamber—offgases are produced that require some secondary treatment, e.g., combustion and filtration.

DAVINCH technology has been used in Japan to destroy 600 Japanese chemical bombs, some containing a mustard agent/lewisite mixture and others containing vomiting agents. The technology has not been used in the United States to destroy non-stockpile chemical munitions.

The size and explosion containment capabilities of versions of the DAVINCH technology are substantially greater than those of the largest treatment technology used in the United States for RCWM (the EDS-2), and its throughput also exceeds that of the EDS-2 by a factor of at least 3. It has demonstrated the ability to destroy over 80 pounds of agent (a lewisite/mustard agent mix in two Japanese Yellow bombs) in a single application and to have destroyed 10.14 pounds of explosive (picric acid) in these bombs.

The DAVINCH technology appears to be safe and effective. The detonation of an externally placed explosive charge allows DAVINCH to be used to open agent-filled containers, inert munitions, and munitions containing energetics in order to access and destroy the agent. DAVINCH is larger and less mobile than the EDS-2, although a transportable version is under development.

Although the specific application of the DAVINCH to meeting future U.S. non-stockpile disposal needs will depend on the nature of the items to be disposed of, DAVINCH technology has potential applicability at those U.S. sites where a temporary facility can be placed and used to dispose of medium to large quantities (hundreds to thousands) of items that either contain chemical agent or are agent contaminated. It is probably not cost-effective as a disposal technology for items unlikely to contain agent, e.g., containers that have been previously burnt out, or for small quantities of smaller chemical items, e.g., bomblets or small-caliber projectiles where the EDS technology would have greater applicability.

Dynasafe Technology

Dynasafe is the trade name for a static kiln manufactured by Dynasafe AB, a Swedish company. The kiln is a near-sphere, armored, dual-walled, high-alloy stainless steel detonation chamber (heated retort) inside a containment structure. The total wall thickness, including a safety layer, is 15 cm. The detonation chamber can operate in a pyrolytic or oxidizing environment. Intact munitions are indirectly heated by electrical resistance elements between the inner and outer walls of the detonation chamber. The munitions are heated to 400°C-600°C, resulting in deflagration, detonation, or burning of the munition's explosive fill. The chemical agent in the munition is destroyed as a result of the shock wave from the detonation, the resulting gas pressure (measured at 10 bars, or 9.87 atmospheres), and the heat within the detonation chamber. No explosive donor charge is used, nor is a reagent needed to neutralize the agent. The kiln operates in a semibatch mode.

Chemical munitions are placed in a cardboard box or carrier, which is transported to the top of the kiln. The boxed munitions are fed into the kiln through two loading chambers, each having its own door. The boxed munitions are dropped onto a heated (500°C-550°C) shrapnel (scrap) bed at the bottom of the detonation chamber. If sufficient energy from energetics in the munition is released, no additional external heating from the electrical resistance elements is required. If the munition does not contain energetics, additional heat can be provided by the electrical resistance elements.

The Dynasafe technology has been demonstrated to be effective in destroying small conventional munitions and explosives, in destroying some chemical agents, and in destroying mustard-agent-filled, explosively configured German grenades. The technology could be viable for disposing of U.S. non-stockpile chemical munitions provided that continued operation at the German GEKA testing facility (ongoing as this report was being prepared) demonstrates its ability to safely and effectively access the agent in German munitions, destroy a variety of chemical agents, and process secondary wastes.

The Dynasafe technology could find application at U.S. sites where fairly large numbers of chemical munitions, such as bomblets, mines, 105-mm projectiles, and 155-mm projectiles, need to be recovered and where effective use could be made of its high throughput. Its limited explosive containment capacity, however, limits it to destroying items containing up to 5 pounds TNT-equivalent, about the same as the EDS-2. This limited capacity also means a Dynasafe operator may not introduce into the detonation chamber high explosive rounds that would exceed the chamber's explosive containment capacity. Even with a 100 percent safety margin—allowing up to 10 pounds TNT-equivalent of explosive loading—the detonation of such rounds could reduce the life of the chamber and, in the worst case, severely damage it.

The Dynasafe technology depends on heat rather than donor charges to detonate energetics within a munition and to access the agent fill. This process is expected to be effective for chemical munitions that contain energetics but may be more problematic for inert chemical munitions if the munition emerges from the detonation chamber intact and if in situ agent destruction cannot be confirmed. If it can be demonstrated that agent destruction does take place regardless of the munition configuration (energetics vs. inert) or the condition of the munition following treatment in the detonation chamber (intact vs. in fragments), then the Dynasafe static kiln can be an effective and flexible technology for destroying large quantities of chemical munitions, within its explosive containment and munition size constraints.

TIER 1 INTERNATIONAL TECHNOLOGIES FOR AGENT-ONLY PROCESSING

Two technologies were identified as Tier 1 international technologies for agent-only processing. These are briefly described below and fully covered in Chapter 5 (with additional information given in Appendix C).

Russian Two-Stage Process: Neutralization with Addition of Bitumen

For destruction of nerve agents, the focus in Russia in recent years has been on a two-stage technology for neutralizing the agent (Stage 1) and adding the neutralent to bitumen to form a stabilized mass (Stage 2) that can be safely stored for indefinitely long periods of time. Procedures have been developed for the nerve agents VX, VR (the Russian version of VX), GB, and GD and for mustard agent.

A facility that will use the two-stage process is being built at Shchuch'ye in Russia to destroy much of the 30,000 metric tons of nerve agent stored there. A pilot facility with a capacity of 500 metric tons per year will be built and then expanded to 1,200 metric tons per year. Joint Russian-U.S. laboratory testing carried out to evaluate the process resulted in its acceptance for the destruction of nerve agents in Russia.

Incineration

Incineration is a well-developed technology that has been shown to be effective for destroying stockpiled chemical weapons. At present, incineration is being used in Germany and the United Kingdom for destroying recovered chemical weapons. The U.S. Army and its contractors have developed very advanced and sophisticated incineration technology for the destruction of the U.S. chemical weapons stockpile.

However, the desired complete conversion of the carbon and hydrogen in organic compounds to carbon dioxide and water is generally not achievable using incineration technology. Instead, trace amounts of compounds such as dioxins, furans, and other products of incomplete combustion can be generated during the combustion process and must be controlled in an offgas treatment system. This characteristic of the incineration processes has been a source of difficulty in gaining public acceptance for this technology, especially from stakeholders in local communities and environmental interest groups.

The baseline incineration process employed by the U.S. Army to destroy stockpiled chemical weapons that are in reasonably good condition is not useful for the destruction of non-stockpile chemical weapons because the deteriorated condition of the latter will not allow their disassembly with the existing equipment. The committee postulates that any use of incineration by the United States in the future for destroying recovered chemical weapons (other than, of course, the use of the currently operating baseline incineration facilities to destroy the U.S. stockpile) would be done only as a last resort in special situations and would be primarily for the destruction of agent stored in bulk containers or recovered from bombs and other weapons.

TIER 2 INTERNATIONAL TECHNOLOGIES FOR MUNITIONS PROCESSING

The committee considered a number of additional technologies but judged them not to be as promising as the Tier 1 technologies previously discussed. These Tier 2 technologies are listed below and are described and discussed in Chapter 6.

The following Tier 2 processes for destroying complete munitions are examined:

- Acid digestion (France),
- Bulk vitrification (United Kingdom), and
- Firing pool (France).

Six Tier 2 processes for destroying only agent from recovered CWM are examined:

- Biological approaches (several countries),
- DSTL electric furnace (United Kingdom),
- Electrochemical oxidation (United Kingdom and United States),
- Photocatalysis (Scotland),
- Plasma arc (Switzerland), and
- Plasmazon (Germany).

OTHER TECHNOLOGIES RELEVANT TO NON-STOCKPILE OPERATIONS

In the course of researching international CWM treatment technologies, the committee also identified and compiled information on technologies used to detect, assess, access, and remediate the contents of large burial sites. These sites have not been thoroughly characterized and their exact contents remain unknown. This effort was not included in the statement of task. However, in early committee meetings, the committee was asked by NSCMP staff to report on the existence of any promising international technologies that it encountered during its information gathering for assessing chemical weapon burial sites and accessing recovered chemical munitions.

DOD is a leader in the research and practice of detecting subsurface munitions and explosives of concern using geophysical processes. Since the mid-1980s, there have been numerous investigation and remediation projects for conventional (high-explosive) munitions and explosives of concern under various DOD programs such as the base realignment and closure program and the formerly used defense sites program. Since that time, geophysical techniques and technologies for the detection of subsurface munitions and explosives of concern have been developed. It is now possible to detect individual or mass buried munitions and explosives of concern, with magnetometry and active geophysical systems being the most common and productive technologies. In addition, DOD has programs supporting research and development in this technical area. However, the technical challenges associated with assessing the contents of large, identified chemical munitions burial sites have not been specifically addressed. The committee's research into foreign technology did not reveal any potential breakthroughs in this area using geophysical sensors.

Some sensing technologies should be investigated further. One is the use of chemical agent detector dogs to locate subsurface buried CWM. The U.S. Bureau of Customs and Border Protection is using chemical detector dogs to detect CWM. These dogs have a detection capability three to five orders of magnitude greater than that of today's best instruments. The committee also found that the United Kingdom plans to conduct tests at Porton Down to determine the effectiveness of chemical agent detector dogs.

There are also some potentially useful agent-sensing technologies that do not rely on biological sensors. These new devices may offer greater simplicity in measurement, rapid analysis, and continuous measurement. One group of new sensors is known as electronic or artificial noses. An array of semiselective, cross-reactive sensors produces a response pattern characteristic of a chemical. The patterns are preprogrammed mathematically so that upon exposure, they can be matched to the chemicals sensed.

Japan is planning to use a telerobotic and automated system for excavating, handling, and disposing of 300,000 to 400,000 World War II-era CWM abandoned by Japan at the large burial site at Haerbaling, Jinlin Province, China. For this project, the Japanese are designing a combination remotely operated and automated excavation system consisting of excavation robots, a device to remove attached soil using pressurized air, and an automated transportation system that will take the removed CWM through a series of cleaning and assessment stations and then finally to a packing station and temporary storage.

KEY FINDINGS AND RECOMMENDATIONS

Finding 4-1. The U.S. Army's EDS, although proven to be safe and effective, has a low throughput rate, is limited in the size of the munitions it can handle, and generates a liquid waste stream that must be disposed of. Consequently, while it will continue to have application for small quantities of munitions, EDS would be expected to have limited applicability to the destruction of the anticipated large quantities and variety of munitions and agent-contaminated items expected to be found at large burial sites in the United States.

Finding 4-2. Detonation-type technologies offer complementary capabilities to the EDS and all have the following characteristics:

- There is no agent neutralization step.
- All are total solutions—that is, they all access the agent, destroy the energetics and agent, and decontaminate the munition bodies.
- All require secondary thermal or catalytic treatment of offgases.
- All have a higher throughput than the EDS and the same or greater explosive containment capability.
- All have been operated safely.

Recommendation 4-1. The U.S. Army should select a detonation-type technology as the method for destroying recovered chemical munitions excavated from a large burial site, although the EDS will continue to have application, especially at small sites. In view of the rapidly evolving development efforts on the three international detonation-type technologies, the U.S. Army should monitor the operations and capabilities of these technologies and collect cost and performance data with the goal of selecting one of them as the primary technology.

Recommendation 4-2. To further the evaluation of detonation-type technologies for non-stockpile applications, the U.S. Army should establish accepted procedures that effectively and efficiently determine the degree of agent destruction or in some other way measure the performance of these processes. The procedures should involve the feeding of complete munitions to the process—that is, munitions containing either agent or a chemical surrogate that is more difficult to destroy than the chemical agent that is most resistant to destruction. Both the degree of agent destruction in the actual detonation event and the degree of agent destruction in the system overall should be determined. Such procedures should be developed with input from all of the relevant stakeholders.

1

Introduction and Background

PURPOSE OF THIS REPORT

National Research Council (NRC) reports have evaluated a wide range of destruction technologies developed or implemented in the United States. The obligation to destroy recovered chemical warfare materiel (RCWM) applies to many other nations. In the last few years, a number of chemical warfare materiel (CWM) destruction technologies have been developed or refined outside the United States. As discussed in Chapter 2, the United States will increasingly be faced with the need to evaluate and determine how to address buried CWM. The time, therefore, is apropos to evaluate the international CWM destruction technologies that could offer complementary capabilities or possibly even replace current equipment. The statement of task with which the committee was charged is as follows:

> The NRC will establish a committee to review and evaluate international technologies for the destruction of non-stockpile chemical materiel. The committee will provide independent scientific and technical evaluations of international systems, facilities, and disposal technologies currently employed or under research and development in countries with inventories of non-stockpile materiel for their treatment and destruction. The committee will compare these technologies with those utilized by the U.S. Army Project Manager for Non-Stockpile Chemical Materiel, in an overall effort to determine and further define state-of-the-art technologies for destruction worldwide of non-stockpile chemical materiel. The committee will:
>
> > Review and evaluate systems and technologies employed or under development in countries with inventories of non-stockpile materiel for the treatment and destruction of non-stockpile munitions, materiel, and secondary waste streams. Such countries include, but are not limited to, France, Germany, Japan, Russia, and the United Kingdom.
> >
> > Compare and contrast foreign disposal technologies, facilities, and/or systems and their present or future potential to be more effective for the overall disposal of specific types of non-stockpile materiel, as compared with corresponding disposal technologies, facilities, and/or systems presently in use by the U.S. Army Project Manager for Non-Stockpile Chemical Materiel. This comparison will include an assessment of technical feasibility, level of maturity, and overall degree of scientific acceptance versus the disposal technologies presently in use by the U.S. Army Project Manager for Non-Stockpile Chemical Materiel, as well as other items or areas detailed below.
>
> As part of this analysis, the committee will:
>
> > Consider implementation and deployment issues related to cost, safety, risk, and protection of the environment of the foreign technologies and systems
> >
> > At this early stage of assessment of systems and technologies, address acceptability to regulators and stakeholders to the extent that the committee judges that significant problematic issues exist or are relevant.
>
> The NRC will deliver its report to the sponsor within 14 months of contract award.

STUDY SCOPE AND STRUCTURE

Scope

This report primarily evaluates technologies for the complete destruction of recovered non-stockpile munitions and, to a lesser degree, technologies more suited to the destruction of recovered non-stockpile chemical agent only. Since many of the sources of information on destruction technologies were also the sources of information on technologies addressing remote detection and accessing of buried CWM, and since it was asked to do so by the U.S. Army as sponsor of the report, the committee collected the latter type of information as well. It did so to the extent that resources were not significantly diverted from the primary purpose of the study. This information is presented in Chapter 7.

The committee reviewed the scientific literature and publicly available reports prepared for governmental entities of the United States and other countries and interviewed, among others, government scientists and engineers in Belgium, Germany, Japan, Switzerland, the United Kingdom, and the United States; staff in the Organisation for the Prohibition of Chemical Weapons; and representatives of vendors of international treatment technologies for the destruction of CWM. These communications included face-to-face meetings, site visits, telephone calls, and exchanges of e-mails. Where data were available, they were evaluated; however, of necessity, comparative evaluations of technologies rely on the professional judgment and experience of the committee members. It must be acknowledged that, in some cases, it was more difficult (or even impossible) to obtain reliable, detailed technical information from foreign government organizations and technology developers than from the U.S. Army and its contractors.

Structure and Tiering of Technologies

The committee determined that it would be beneficial to organize its evaluation activities in a manner that would enable it to concentrate its efforts on international technologies that appeared to be the most promising for use in non-stockpile operations. This was accomplished by separating the technologies into two groups: (1) those applicable for destroying an entire non-stockpile recovered munition, including agent, energetics, and other materials, such as the munition casing, and (2) those for destroying agent only. The technologies were further categorized in a tiered matrix. The most promising (Tier 1) technologies for munitions processing and agent-only processing are evaluated in Chapters 4 and 5, respectively. Other (Tier 2) technologies for both munitions processing and agent-only processing are described in Chapter 6.

Technologies assigned to Tier 1 were those that appeared to have a good level of maturity and to possess capabilities required by the U.S. Army's Non-Stockpile Chemical Materiel Project (NSCMP), as determined by the committee after discussions between the committee and NSCMP staff. Tier 2 technologies are of two basic types. The first type are those technologies with potential applicability to NSCMP projects but that are still at an early stage of development for such applications. The second type of Tier 2 technologies are those that have been tried out in operations for destroying recovered chemical weapons materiel but that have not, for various reasons, proven to be satisfactory. The evaluation factors analysis described in Chapter 3 was applied only to the Tier 1 technologies.

The international Chemical Weapons Convention (CWC) treaty deadline for destroying CWM that have already been recovered is April 29, 2007. This date is near, so the committee focused on evaluating international destruction technologies that could satisfy operational requirements for destroying non-stockpile CWM that has yet to be recovered. In particular, the committee was interested in examining technologies that could be effectively implemented at sites where large quantities of buried materiel could be expected and where, consequently, higher throughputs for destruction operations might be achieved than with current NSCMP equipment. The committee considered the applicability of international technologies for sites where only single or a few items might be recovered. With the exception of offgas treatment technologies, the report does not specifically address separate technologies for treating secondary waste.

Report Organization

Chapter 1 provides background information on the NSCMP and describes the purpose and approach of the report. Chapter 2 summarizes the factual, regulatory, and other characteristics of buried CWM sites to set the stage for evaluating the international technologies examined in this report.

Chapter 3 explains the criteria used to evaluate the international technologies. Chapter 4 applies these criteria to the Tier 1 treatment technologies for complete destruction of recovered CWM munitions, and Chapter 5 applies the criteria to Tier 1 technologies for treatment of recovered chemical agent. Chapter 6 describes and comments on technologies the committee assigned to Tier 2 status. Chapter 7 reports on technologies relating to the remote detection and accessing of buried CWM. That information was gathered by the committee in the course of its research on the primary treatment technologies.

U.S. NON-STOCKPILE PROGRAM

Chemical Demilitarization Overview

The elimination of the extensive inventory of weapons containing chemical agents and of chemical agent in bulk that has been maintained by the United States has been in progress since the early 1990s. This inventory, known as the chemical weapons stockpile, or simply "stockpile," was developed during World War II and in the following decades. Since then, it was or continues to be maintained at eight storage sites in the continental United States and on Johnston Atoll in the Pacific Ocean, southwest of Hawaii.

Destruction operations on Johnston Atoll using the U.S. Army's baseline incineration system were completed in 2000. Destruction operations using this technology at four of the continental U.S. storage sites are currently in progress. At the other four storage sites, technology based on the use of hydrolysis for destruction of agent (and, where applicable, energetic material) has been employed or is planned. At one of the latter sites (Aberdeen, Maryland), where mustard agent HD was stored in bulk, destruction operations have already been completed. Destruction of the entire U.S. stockpile was slated to be complete by April 29, 2012. This date represents the 5-year extension allowable under the terms of the CWC treaty, which was adopted on April 29, 1997, and called for all declared chemical weapons to be destroyed in 10 years (2007), with the possibility of a 5-year extension. However, current estimates are that only about two-thirds of the 31,500-ton original stockpile will be destroyed by 2012.

In addition to the stockpile, U.S. law and international treaties recognize another category of CWM, which in the United States is designated as non-stockpile materiel. This category includes materiel that was buried on current and former military sites, some of which is now being recovered as the land is remediated. Non-stockpile munitions have been found in a variety of conditions owing to their exposure to uncontrolled environments and are generally not amenable to processing by the reverse assembly (disassembly) process used in the baseline stockpile incineration program. Non-stockpile materiel that had been recovered and was in storage prior to the ratification of the CWC is required to be destroyed by April 29, 2007. Within the U.S. Army, the NSCMP has been charged with a number of mission tasks to accomplish this under the direction of the Chemical Materials Agency, which is charged with the overall management of U.S. chemical demilitarization activities. These tasks, which include the demolition of former production facilities (such as those at Aberdeen Proving Ground–Edgewood Area, Maryland, and Newport Chemical Depot, Newport, Indiana) and the destruction of recovered non-stockpile materiel that has been in storage at stockpile locations (such as at Pine Bluff Arsenal, Arkansas), are well under way and are expected to be completed by the treaty deadline.

However, a large quantity of non-stockpile materiel has not yet been recovered; it is discussed later in this chapter and in Chapter 2. This materiel is not subject to CWC requirements until it is recovered. Once the recovery of such materiel is declared, however, the CWC calls for it to be destroyed "as soon as possible." This report explores the technology options available in other countries that the U.S. Army might wish to consider using for these future operations.

Chemical Weapons Convention

The United States ratified the CWC in 1997.[1] The CWC prohibits the use of chemical weapons and mandates the elimination of existing declared stockpiles by April 29, 2007, but allows the deadline to be extended to 2012, a provision that has been or is likely to be invoked, at least for stockpiled chemical weapons, by some of the parties to the CWC.

In the CWC framework, "old chemical weapons" are chemical weapons that (1) were produced before 1925 or (2) were produced in the period between 1925 and 1946 and have deteriorated to such extent that they can no longer be used as chemical weapons. "Abandoned chemical weapons" are chemical weapons, including old chemical weapons, abandoned by a state after January 1, 1925, on the territory of another state without the consent of the latter. Each party to the CWC must declare whether chemical weapons have been abandoned on its territory, and the party that abandoned such weapons must also declare that it did so. The party that abandoned the chemical weapons is obligated to destroy them. Although there are a significant number of abandoned chemical weapons, the exact location of most of the abandoned weapons is not public information.

These two categories do not impact the obligation to destroy such weapons, since the CWC requires both old chemical weapons and abandoned chemical weapons to be declared (whether on a member nation's own territory or the territory of another) and destroyed within the CWC treaty deadline if they are unearthed prior to the deadline (Pearson and Magee, 2002). However, because old and abandoned CWM might also be found after the CWC deadline, their disposal will likely continue after 2007. The CWC requires abandoned weapons to be destroyed as toxic waste in accordance with the national regulations of the country in which the weapons reside.

More generally, the term "buried chemical warfare materiel" refers to any CWM buried prior to January 1, 1977, or dumped at sea prior to January 1, 1985. Any CWM discovered and recovered after the initial declaration that was required by the CWC treaty must be destroyed (including formerly buried CWM). However, the CWC allows a member nation to exclude CWM buried on its territory before January 1, 1977, or disposed of at sea before January 1, 1985, as long as the materiel remains buried (U.S. Army, 1996). Thus, the CWC does not require buried CWM to be declared or destroyed as long as the materiel remains buried. At the time this report was prepared, CWM dumped at sea did not fall under the authority of the PMNSCM and were therefore not directly addressed by the committee.

In the United States, when buried CWM are removed from their burial site (i.e., when they become RCWM), they must be identified, classified, declared, and disposed of in accordance with CWC, Environmental Protection Agency (EPA), and state environmental regulations (U.S. Army, 1996, 2001a, 2004). The CWC allows some negotiation of the timetable for the disposal of CWM declared after the treaty's entry into force, although generally it should be "destroyed as soon as possible."[2]

[1]The Convention on the Prohibition of the Development, Production, Stockpiling and Use of Chemical Weapons and on their Destruction can be found at <http://www.opcw.org/>; it is the basis for this section of the report.

[2]Information derived from a meeting between representatives of the Organisation for the Prohibition of Chemical Weapons and the members of the committee, The Hague, The Netherlands, January 18, 2006.

Types of Non-Stockpile Items

Non-stockpile chemical warfare materiel (NSCWM) is far more diverse than stockpile CWM: For example, it contains U.S. unitary munitions and accessories dating back to World War I, binary munitions, and foreign munitions brought back to the United States after World War II. There is a greater variety of chemical agents in NSCWM than in stockpile materiel (including blister agents, nerve agents, blood agents, and choking agents, as well as militarized industrial chemicals). Energetics found in chemical munitions include aromatic nitro compounds such as trinitrotoluene (TNT), aromatic nitramines such as tetryl, heterocyclic nitramines such as cyclotrimethylenetrinitramine (RDX), high-melting explosive cyclotetramethylene-tetranitramine (HMX), and nitrate esters used in propellants (e.g., nitrocellulose and nitroglycerine). The most commonly encountered energetics are tetryl, TNT, and composition B (60 percent RDX, 39 percent TNT, 1 percent wax).[3] The condition of the NSCWM is also much more variable than that of the stockpile, especially for items that have severely deteriorated after being buried for decades (NRC, 2002).

Chemical agent identification sets (CAIS), which were also disposed of by burial and are frequently found, are also NSCWM.[4] CAIS are kits consisting of small vials or ampules of diluted or full-strength chemical agents that were used by the U.S. Army for training troops to recognize the odor and the effects of CWM.

Appendix A shows the inventory of recovered non-stockpile items that have been stored at various locations awaiting treatment and disposal, which is scheduled to be completed prior to the 2007 CWC treaty deadline using currently available NSCMP equipment. Except for the listed binary agent precursors (which postdate the period when the Army practiced burial of non-stockpile chemical warfare materiel), the tables in Appendix A exemplify the great variety of items that could be encountered during future recovery operations.

Scope of Buried Non-Stockpile Chemical Weapons Materiel

As of 1996, the U.S. Army had located 168 potential CWM burial sites at 63 locations (primarily current or former military facilities) in 31 states, the U.S. Virgin Islands, and the District of Columbia (U.S. Army, 1996).[5] They include sites with CAIS only, sites with small quantities of materiel with and without associated explosives, and sites with large quantities of materiel with and without explosives. The majority of the sites involve small quantities of materiel (NRC, 2002). Based on its 1996 survey, DOD estimated the cost of disposing of buried CWM at more than $11 billion (DOD, 2003). However, the committee has not reviewed these estimates and expects that the actual cost will depend on, among other factors, the number of large burial sites, the degree to which active removal and destruction (as opposed to containment in place) is chosen as the remedy, and the number of buried CWM found in residential areas.

The universe of buried CWM includes several sites where it is known that large numbers (in the thousands) of buried CWM are located—Redstone Arsenal, Alabama, which reportedly contains a mustard agent site and disposal trenches; Rocky Mountain Arsenal, Colorado, which has large numbers of buried CWM in Basin A; Aberdeen Proving Ground, Maryland, where there are large quantities in at least the Old and New "O" Field landfills; and Deseret Chemical Depot, Utah (formerly Tooele Army Depot South Area), particularly in solid waste management units 1 and 25 (U.S. Army, 1996).[6] There also may be medium to large numbers of buried CWM at Pueblo, Colorado; Black Hills, South Dakota; Newport, Indiana; and Schofield Barracks, Hawaii.[7] (See Table 1-1.)

The characterization of a site as having large, medium, or small numbers of buried CWM in this report is qualitative. The prioritization model described in Chapter 2 is intended to more explicitly quantify the amount of CWM and should increase the priority of sites with large numbers of buried CWM that could one day pose a significant risk.

An update of the 1996 survey is undergoing internal DOD review and was not available in time to inform this report. Furthermore, since new information periodically becomes available, more significant buried CWM sites could be uncovered in the United States or abroad.[8]

Ultimately, the cost of cleanup of buried CWM is not known or knowable precisely because the number of buried CWMs, the future use of the land in which they are buried, and the remedy that will be selected are not yet known.

EXISTING NON-STOCKPILE DESTRUCTION TECHNOLOGIES

This section provides basic descriptions of NSCMP equipment currently used to destroy RCWM. A number of NRC committees have reviewed and evaluated these technologies in depth in previous reports (NRC, 2001, 2002, 2004). In addition to the descriptions presented here, technical information on these NSCMP systems will appear throughout

[3]Stone & Webster information paper briefed to an NRC committee on October 14, 1999.

[4]See NRC (1999) for additional details on CAIS.

[5]The 1996 report was a second, updated version of the 1993 survey that was required by Public Law 102-484, the National Defense Authorization Act for Fiscal Year 1993, section 176.

[6]William R. Brankowitz, Deputy PMNSCM, at a meeting of the committee, Washington, D.C., November 29, 2005.

[7]Information provided to the committee by the NSCMP.

[8]No attempt was made in this report to address recovery or treatment of chemical weapons materiel that was disposed of in the ocean in the decades following World War II. However, the technologies described in this report could conceivably be used for such materials recovered in the future.

TABLE 1-1 Examples of Known or Potential Large Sites of Buried CWM Identified by the U.S. Army

Type of Site	Name of Site	Description	Issues	Possible Approaches to CWM Disposal
Live-fire sites	Camp Sibert, Ala.	Firing range for live CWM, 4.2-in. mortars	Likely to contain dud-fired CWM (potentially shock sensitive)	1. Detect subsurface geophysical anomalies 2. Hand excavate anomalies in a containment structure 3. Dispose of CWM in EDS (low volume, less than 100 CWM expected)
	Schofield Barracks, Hawaii	CWM live-fire range	Likely to contain dud-fired CWM (potentially shock sensitive)	1. Detect subsurface geophysical anomalies 2. Hand excavate anomalies in a containment structure 3. Dispose of CWM in EDS (volume unknown)
Large burial sites	Pueblo, Colo.	Suspected to contain buried M70 bombs in burn/burial pits	Site will be closed under Base Realignment and Closure	1. Excavate pits (using appropriate methods, either manual or robotic) 2. Identify potential CWM as it is uncovered 3. Assess potential CWM using x-ray and portable isotopic neutron spectroscopy 4. Dispose of CWM in appropriate system(s) staged at the site
	Tooele/Deseret Chemical Depot, Utah	Approx. 25 pits where CWM bombs and 4.2-in. mortars were burned and buried	Many CWM munitions at this site are expected to be empty due to burn/burial process	
	Redstone Arsenal, Ala.	50-60 pits, 100 yards long × 20 ft wide; used by technical escort unit to dispose of U.S. and foreign CWM after World War II; CWM stacked, burned, and buried	Large quantity of CWM expected, but most are likely to be empty due to burn/ burial process; significant amount of foreign CWM is expected	
	Black Hills, S.D.	Large CWM storage facility in WWII	Potential large burial site	
Large burial sites with records of decision (RODs)	Old "O" Field, Aberdeen Proving Ground, Md.	Multiple pits, contain U.S. and foreign CWM in various states and conditions	Extremely dangerous sites, spontaneous deflagrations observed presumably due to mixing of incompatible chemicals and explosives	Site is capped under a regulatory ROD. Remedy is a bentonite barrier, sand cap, and a pump-and-treat system; remedy appears to be working well and is reviewed periodically for effectiveness; ROD is reviewed periodically and further remedial action is possible
	Rocky Mountain Arsenal, Basin A, Colo.	Large (1 sq. mi) site, numerous burial trenches suspected to contain CWM, vehicles, scrap, equipment, etc.	Site used for disposal by commercial chemical manufacturers making pesticides	Site is stabilized under an ROD; barrier wall and pump-and-treat system installed; ROD is reviewed periodically and further remedial action is possible

SOURCE: William R. Brankowitz, Deputy PMNSCM, presentation to the committee on November 29, 2005.

Chapters 4 through 6 as the capabilities of the international destruction technologies are discussed and compared with the capabilities of the current suite of NSCMP equipment, with particular emphasis on the EDS.

Explosive Destruction System

The EDS is capable of treating munitions regardless of whether or not they are energetically configured.[9] The heart of the EDS system is an explosion containment vessel mounted on a flatbed trailer (see Figure 1-1). The EDS Phase 1 unit (EDS-1) has an inside diameter of 20 inches (51 cm), is 36 inches (91 cm) long, and can process munitions containing up to 1 pound TNT-equivalent of explosives. The EDS Phase 2 (EDS-2) has an inside diameter of 28 inches (71 cm) and a length of 56 inches (142 cm) and is designed to handle munitions containing up to 3 pounds TNT-equivalent of explosives, with occasional uses up to 5 pounds TNT-equivalent of explosives. The EDS is intended for use with World War I and World War II vintage CWM produced before 1945. In general, post-World War II munitions have bursters that exceed the capacity of the system.

[9]Unless otherwise noted, material from this section was drawn from NRC (2001, 2002).

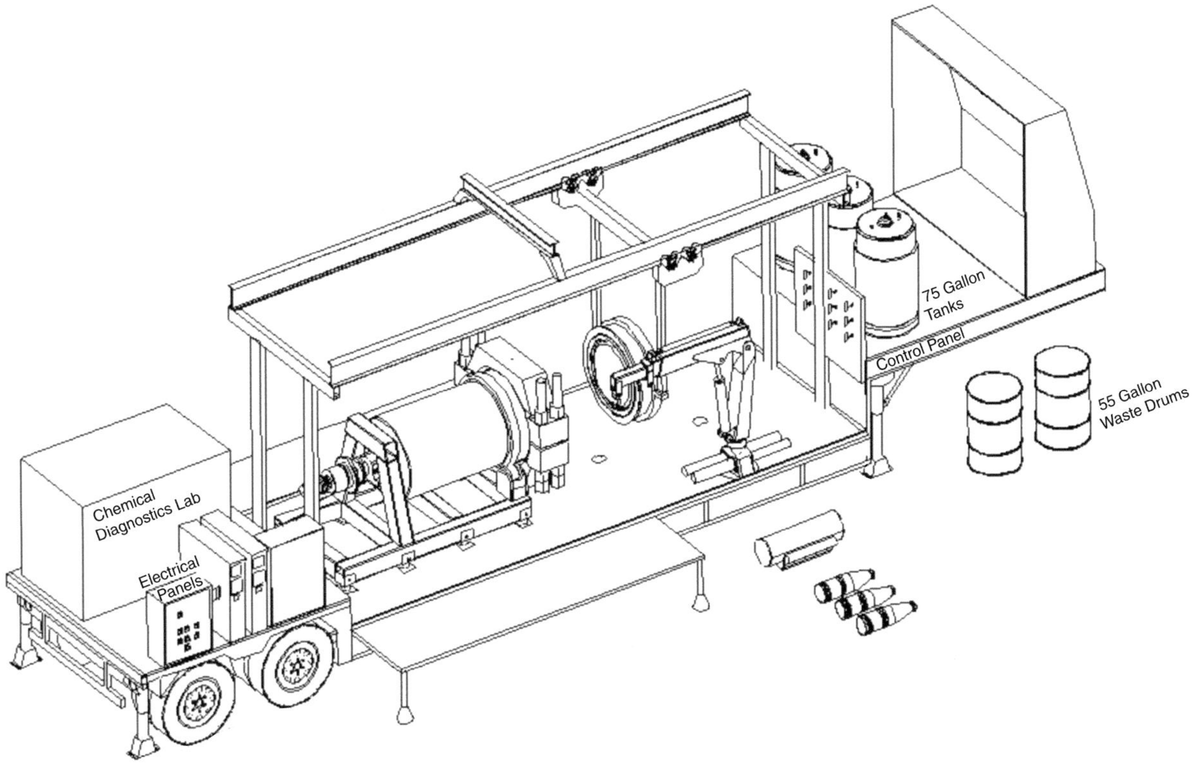

FIGURE 1-1 Diagram of EDS-2. SOURCE: Tom Cain, Mitretek, presentation to the committee on February 21, 2006.

Large safety factors have been built into the design of the EDS vessel and the procedures for its operation. The mechanical integrity of the vessel was evaluated by Sandia National Laboratories using a combination of small-scale failure analysis tests and computer simulations. This evaluation indicated that the EDS-l containment vessel could withstand several thousand detonations with more than 1 pound of explosive, providing a significant margin of safety for a system with an intended life of 500 detonations (SNL, 2000).

The EDS uses explosive shaped charges to access the agent in a munition and to destroy any energetics in the munition, with both processes taking place in a sealed explosion containment vessel. After detonation of the shaped charges, reagents appropriate to the agent to be neutralized are pumped into the vessel and the vessel contents are mixed until the treatment goal has been attained. After the concentration of chemical agent falls below the treatment goal (as determined by sampling the contents of the chamber), the liquid waste solution is transferred out of the chamber into a waste drum. The drummed EDS liquid waste is treated further at a commercial hazardous waste treatment, storage, and disposal facility (TSDF).

The EDS-1 can be driven or flown on a C-130 aircraft to a site where non-stockpile chemical materiel is discovered or recovered materiel has been stored. The EDS-2 can be driven but has not been evaluated for air transport requirements. The EDS is considered the U.S. Army's transportable system of choice for treatment of small quantities of non-stockpile munitions.[10] It has an excellent record of regulatory compliance and public acceptance during the multiple deployments that have taken place around the nation, including the Spring Valley, Washington, D.C., cleanup operation, which took place in the middle of a residential neighborhood. Multiple EDS units are currently in use at Pine Bluff Arsenal in Arkansas to destroy recovered non-stockpile munitions that have been in storage there.

Rapid Response System

The RRS is a system for the treatment of significant quantities of recovered CAIS, of which approximately 110,000 were produced in various configurations from 1928 through 1969.[11] A CAIS can be a sealed glass tube containing agent, a glass bottle containing agent, or a glass jar containing agent

[10]Sites with a large number (that is, hundreds or thousands) of non-stockpile items were originally expected to be served by semipermanent treatment facilities, although a plan was implemented that replaced the Pine Bluff Non-Stockpile Facility with multiple EDS units for the processing of hundreds of recovered munitions (see NRC, 2004).

[11]Unless otherwise noted, the information for this section is derived from NRC (2001, 2002).

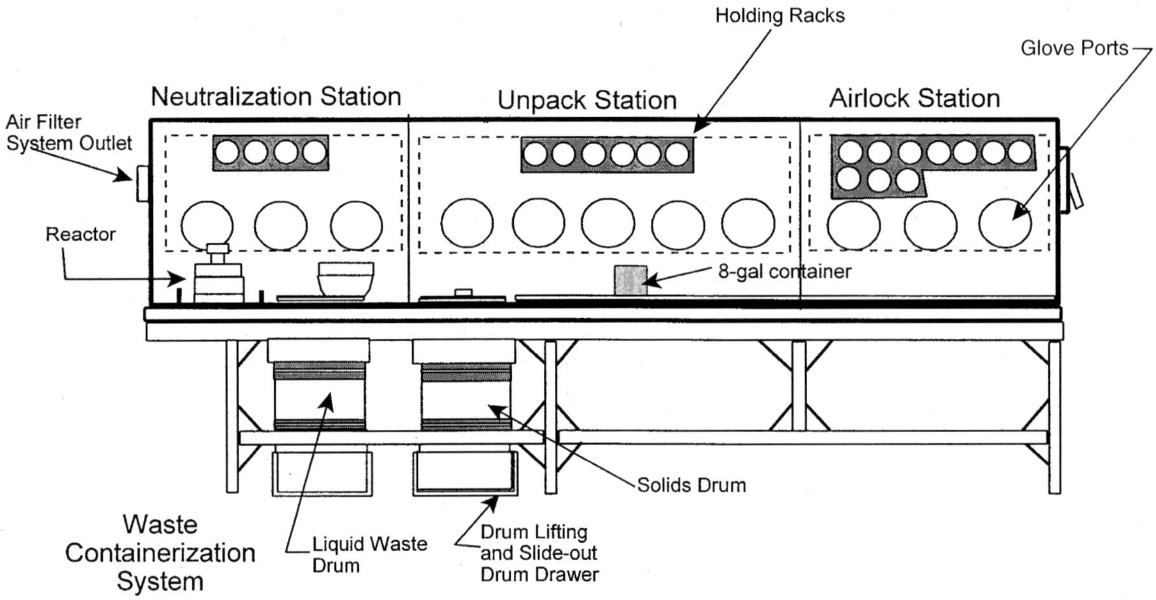

FIGURE 1-2 Diagram of RRS operations trailer. SOURCE: U.S. Army, 2001a.

adsorbed onto granular charcoal for use as a sniff set. The RRS uses a variety of chemical neutralization techniques to destroy agent, the technique depending on the agent.

A line drawing of the RRS is shown in Figure 1-2. The RRS occupies three trailers. The operations trailer contains a series of linked glove boxes equipped to remove CAIS vials and bottles from their packages, identify their contents, and treat those that contain chemical agents (CAIS containing industrial chemicals are segregated and repackaged for off-site commercial disposal). The glass containers are then crushed in a reactor containing a chemical formulation that rapidly neutralizes the chemical agent. The contents of the reactor, including reagent, solvents, agent degradation products, and glass fragments, are transferred to sealed containers for disposal at a commercial hazardous waste TSDF. The support trailer contains spare equipment and supplies. The utility trailer carries electrical generators to allow the system to operate without commercial or host power when needed. For a more complete description of the RRS equipment and operations, see *Rapid Response System Test Report* (U.S. Army, 2001b).

The RRS can be either driven to or flown to locations where CAIS have been recovered. Transporting by air requires the use of two C-17 aircraft (one for the RRS operations and utility trailers and one for transporters, a supply trailer, and a mobile analytical support laboratory). The RRS can treat one PIG[12] of CAIS per day.

[12]A PIG is a metal canister with packing material designed to protect CAIS during transport.

The RRS has been successfully deployed to a number of sites around the nation. The first deployment was to Fort Richardson, Alaska, in July 2003. The RRS is currently destroying CAIS items at Pine Bluff Arsenal (PBA), having destroyed 1,000 of 5,000 CAIS items scheduled for destruction at PBA as of November 2005. The PBA deployment is scheduled to end by February 2007.[13]

Single CAIS Accessing and Neutralization System

The SCANS reactor is a small, disposable container used to access and treat individual CAIS vials or bottles containing chemical agents (see Figure 1-3).[14] Its process chemistry is similar to that of the RRS neutralization. It is intended for use only where a limited number (80 or fewer) of loose CAIS vials or glass bottles are recovered. Because SCANS does not have the glove box necessary to open a CAIS PIG safely, it could not be used for destruction of a CAIS PIG. The SCANS is a hand-held device. It requires neither the elaborate system of trailers that supports the RRS nor its large operating crew. It is a relatively inexpensive destruction system.

The SCANS is used in conjunction with an analytical system such as a portable Raman spectrometer or a portable isotopic neutron spectrometer to identify the agent inside a vial or bottle so the correct reagent can be selected to neutralize it. A 4-liter bottle of reagent is placed in the reactor case,

[13]See Rapid Response System at Pine Bluff Arsenal, <http://www.cma.army.mil/docviewerframe.aspx?docid=003671063>.

[14]Unless otherwise noted, information is drawn from NRC (2002).

FIGURE 1-3 Photograph of SCANS. SOURCE: Tom Cain, Mitretek, presentation to the committee on February 21, 2006.

along with a single CAIS item. The reactor is sealed and a breaker rod manually driven through the reactor, breaking the containers holding the CAIS agent and reagent. The agent mixes and reacts with the reagent to form a neutralent solution. The neutralent-containing reactor is then shipped to a permitted hazardous waste TSDF.

Neutralization and Hydrolysis

Other technologies currently used in non-stockpile operations, and also in stockpile operations, are neutralization and hydrolysis for destruction of chemical agent. This report employs the term "neutralization" for the use of an organic reagent to destroy agent and "hydrolysis" for the use of an aqueous reagent to destroy agent.[15] The use of nonaqueous neutralization in the EDS, RRS, and SCANS was noted earlier in this chapter. As previously mentioned, the baseline incineration system is being used to destroy the U.S. stockpile of obsolete chemical agents and munitions at four continental U.S. storage sites. The U.S. Army Chemical Materials Agency has used or is planning to use hydrolysis technologies at the four remaining storage sites in the continental United States. Hydrolysis with hot water at 90°C (194°F) was used to destroy the stockpile of bulk mustard agent (HD) stored in ton containers at the Aberdeen, Maryland, site, with a destruction efficiency for mustard agent of more than 99.9999 percent. The resulting hydrolysate was sent offsite for biotreatment. Hydrolysis with a caustic NaOH solution at 90°C (194°F) will be used to destroy the bulk stockpile of VX nerve agent at the Newport, Indiana, site. At the Pueblo, Colorado, site, where mustard agent (HD and HT) is contained in nearly 8,000 projectiles, the agent will be removed from the assembled weapons and then hydrolyzed with hot water at 90°C (194°F) (NRC, 2005a). As currently planned, the hydrolysate will be biotreated at the site. At the Blue Grass site, both nerve agents (GB and VX) and mustard agent (H) are contained in a variety of munitions. Hydrolysis-based technology similar to that described above will be used (NRC, 2005b). See Table 1-2 for a list of some process parameters for neutralization of the agents at the Blue Grass, Kentucky, site.

The chemistry of the hydrolysis reactions has been extensively studied (Yang et al., 1992; Yang, 1995). A good agent-by-agent summary of hydrolysis/neutralization is given by Pearson and Magee (2002). As shown in Table 5-1 in Chapter 5, neutralization and hydrolysis of agent have been used extensively in past operations in the United States and other countries.

[15]The terms "neutralization" and "hydrolysis" are often used interchangeably in the literature on chemical agent demilitarization. Hydrolysis is the more appropriate term from a chemical process perspective. Neutralization is more in keeping with the notion of "to neutralize" and thereby render innocuous. It may be found in the literature to refer to hydrolysis in either aqueous or nonaqueous media.

TABLE 1-2 Agent Neutralization Parameters for the Blue Grass Chemical Agent Destruction Pilot Plant

	Agent		
	GB	VX	H
Agent process feed concentrations			
Agent (wt %)	7.5	16.6	8.6
Caustic (wt %) (from 50% NaOH)	11.34	17.44	Added after to adjust pH to 10-12
Water (wt %)	81.16	65.96	91.4
System parameters and performance specifications			
Operating temperature (°F)	140	194	194
Peak rate baseline (units per hour)	40 GB rockets 15 GB projectiles	48 VX rockets 26 H projectiles	26 H projectiles
Peak rate (lb agent/day)[a]	15,540	15,379	7,301
Total time per batch (min)	168	516	243
Throughput (batches per reactor per day)	6	2.5	2.5
Total weight of agent to be destroyed (tons)	305.7	127.2	90.63
Maximum agent concentration to achieve 99.9999 percent destruction efficiency (ppb)	75	160	85

[a]The peak rate is the maximum expected rate during a campaign. Normal operating rates will be lower. Peak rate agent volumes have been multiplied by a safety factor of 1.25 for the purpose of sizing the tanks and other critical materials handling equipment. The total number of batches per day is less than what can be processed in 24 hours, again providing a design safety margin.

SOURCE: Bechtel-Parsons, 2004; NRC, 2005b.

REFERENCES

Bechtel-Parsons. 2004. System Design Description (SDD) for Agent Collection and Neutralization, Blue Grass Chemical Agent Destruction Pilot Plant (BGCAPP) Project, Rev. A, July 26. Aberdeen Proving Ground, Md.: Program Manager for Assembled Chemical Weapons Alternatives.

Department of Defense (DOD). 2003. The Chemical Demilitarization Program: Increased Costs for Stockpile and Non-Stockpile Chemical Materiel Disposal Programs (D-2003-128), September 4. Arlington, Va.: Department of Defense Inspector General.

NRC (National Research Council). 1999. Disposal of Chemical Agent Identification Sets. Washington, D.C.: National Academy Press.

NRC. 2001. Evaluation of Alternative Technologies for Disposal of Liquid Wastes from the Explosive Destruction System. Washington, D.C.: National Academy Press.

NRC. 2002. Systems and Technologies for the Treatment of Non-Stockpile Chemical Warfare Materiel. Washington, D.C.: National Academy Press.

NRC. 2004. Assessment of the Army Plan for the Pine Bluff Non-Stockpile Facility. Washington, D.C.: The National Academies Press.

NRC. 2005a. Interim Design Assessment for the Pueblo Chemical Agent Destruction Pilot Plant. Washington, D.C.: The National Academies Press.

NRC. 2005b. Interim Design Assessment for the Blue Grass Chemical Agent Destruction Pilot Plant. Washington, D.C.: The National Academies Press.

Pearson, G.S., and R.S. Magee. 2002. "Critical evaluation of proven chemical weapon destruction technologies (IUPAC Technical Report)." Pure and Applied Chemistry 74(2): 187-316.

SNL (Sandia National Laboratories). 2000. Explosive Destruction System Fatigue/Life Cycle Analysis. Albuquerque, N.M.: Sandia National Laboratories.

U.S. Army. 1996. Survey and Analysis Report. 2nd ed., December. Aberdeen Proving Ground, Md.: Program Manager for Chemical Demilitarization.

U.S. Army. 2001a. Final Programmatic Environmental Impact Statement, Transportable Treatment Systems for Non-Stockpile Chemical Warfare Materiel, February. Aberdeen Proving Ground, Md.: Product Manager for Non-Stockpile Chemical Materiel.

U.S. Army. 2001b. Rapid Response System Test Report. Final. October. Aberdeen Proving Ground, Md.: Program Manager for Chemical Demilitarization.

U.S. Army. 2004. Recovered Chemical Warfare Materiel (RCWM) Response Process, EP 75-1-3, November 30. Available online at <http://www.usace.army.mil/usace-docs/eng-pamphlets/ep75-1-3/entire.pdf>. Last accessed February 14, 2006.

Yang, Y-C. 1995. "Chemical reactions for neutralizing chemical warfare agents." Chemical Industry 9: 334-337.

Yang, Y-C., J.A. Baker, and J.R. Ward. 1992. "Decontamination of chemical warfare agents." Chemical Reviews 92(8): 1729-1743.

2

Issues Bearing on Sites Containing Large Amounts of Buried Chemical Weapons Materiel

INTRODUCTION

Currently, the primary mission of the Non-Stockpile Chemical Materiel Project (NSCMP) includes the destruction of chemical warfare materiel (CWM) already recovered and not included in the stockpile (U.S. Army, 2004a). However, there are several large CWM burial sites within the United States. Congress has mandated that DOD prioritize the cleanup of munition sites (including those that contain buried CWM) and establish a timetable for implementing investigation, evaluation, and cleanup of these sites. As a result, in the near future, DOD will evaluate buried CWM sites. The decision on whether to remove buried CWM at a particular site or contain it in place depends on the degree of risk presented by the buried CWM, the feasibility and cost of such efforts, and whether Congress imposes a direct statutory mandate. If a decision were to be made to remove buried CWM from the ground at such large sites, the recovered CWM would have to be destroyed. This would in essence represent a new DOD cleanup initiative. However, regardless of the ultimate remedy selected, these sites must be investigated and evaluated to determine whether removal is appropriate pursuant to existing DOD mandated programs. As a result, the evaluation of most of the international technologies reviewed in this report was carried out with buried CWM in mind, potentially from these large sites.

To understand the context in which the CWM destruction technologies will be evaluated, this chapter briefly summarizes the regulatory framework in the United States for CWM recovery and destruction, reviews several key issues pertaining to CWM recovery and destruction, and addresses public involvement.

U.S. REGULATORY FRAMEWORK GOVERNING BURIED CWM

In the United States, the recovery and destruction of buried CWM is governed by overarching hazardous waste laws (U.S. Army, 2001) and DOD munition cleanup programs (P.L. 107-107; U.S. Army, 2001, 2003a, p. 6;[1] Federal Register, 2005[2]). DOD conducts munitions responses and other hazardous waste cleanups in accordance with the Comprehensive Environmental Response, Compensation, and Liability Act (CERCLA) and the Resource Conservation and Recovery Act (RCRA) (Federal Register, 2005; U.S. Army 2004b).[3] In some cases, permits or other kinds of regulatory approvals under CERCLA, RCRA, or other environmental laws would be needed. Generally, and in accordance with CERCLA and RCRA, federal and state environmental regulatory agencies select a remedy after balancing the degree of protection of human health and the environment with other factors, such as long-term effectiveness, permanence, toxicology, mobility, volume reduction through treatment, short-term effectiveness, and the preferences of the public (EPA, 1997). Regardless of whether a CERCLA or RCRA cleanup program applies to buried CWM or whether permits or other forms of approval are required, regulatory acceptance of the chosen approach is legally necessary, and public acceptance must by law be considered in the regulatory decision making. However, in a democratic society, public acceptance is important and generally necessary over the long term as a practical matter. For example, despite the fact that many experts have concluded that incineration is safe and consistent with the regulatory requirements applicable to industrial chemicals, Congress directed the Army to evaluate disposal methods other than incineration. Similarly, public

[1]This document requires the Army to manage its cleanup program "under unified vision and overarching strategy to remedy [any] inefficient organizational divide."

[2]This document explains that CWM is included in the definition of munitions.

[3]Superfund is the commonly used term for the Comprehensive Environmental Response, Compensation, and Liability Act (CERCLA, 42 U.S.C., 9601 et seq.), Executive Order (E.O.) 12580, Superfund Implementation (January 23, 1986), and E.O. 13016 Superfund Amendments (August 28, 1996).

concern may be a factor in the decision making of regulators and even DOD. In summary, community members can influence and sometimes take actions that lead to remedial alternatives they oppose being rejected, but they do not have the authority to veto otherwise scientifically acceptable alternatives.

The National Defense Authorization Act for Fiscal Year 2002 required DOD to develop a comprehensive plan for addressing the remediation of such munitions at defense sites and to assess the funding required and the period of time over which such funding will be required (P.L. 107-107; Federal Register, 2003). DOD's Munitions Response Site Prioritization Protocol is designed to assign priority to each site based on its overall criteria. Part of the prioritization methodology considers the amount of CWM that may be contained in the munition site, its likelihood of dispersal, and the condition of the munition (Federal Register, 2005). As a result, the DOD munition sites that contain significant amounts of CWM will typically be ranked higher than a site without CWM. In addition, state and local pressures may lead to increased emphasis at some sites.

Given the nature of the CWM and the public's concern about these items, as a practical matter, once CWM is discovered in residential areas or at any location not controlled by the federal government, it is likely to be deemed a significant risk to human health pursuant to federal or state environmental statutory authorities.

KEY ISSUES PERTAINING TO CWM RECOVERY AND DESTRUCTION

Rate of Munitions Recovery and Destruction

The selection of destruction and disposal technology to be used in the future may depend, in part, on the volume and rate at which buried CWM are recovered. For example, if a small number of buried CWM are recovered each year, the existing NSCMP capabilities (e.g., the rapid response system and the explosive destruction system) might be sufficient. However, if Congress or federal or state environmental regulators were to require DOD to remove and destroy a large number of buried CWM, or if DOD's comprehensive munitions cleanup plan adopted a policy of removing and destroying buried CWM as rapidly as possible, technologies with high throughput rates would be preferable.

Criteria for Determining Whether Buried CWM Are Recovered

DOD's prioritization methodology determines the sequence in which munition sites should be investigated and evaluated, not whether or what type of remedial action will be taken at any given site. For example, a site could have a high priority, but the remedy selected might be "leave in place with long-term stewardship." The decision on whether to contain in place or remove and treat buried CWM is governed by the same remedy selection criteria that govern hazardous waste sites.[4]

Not all remedial actions involve excavation and treatment. Generally, engineering controls, such as containment, and restrictions on use and access, as well as continued monitoring, can be used for waste that poses a relatively low long-term threat or where treatment is impracticable.[5] For example, cumulatively, through most of FY 2002, EPA selected containment in 43 percent of the CERCLA source control Records of Decision (EPA, 2004). Thus, if the history of hazardous waste cleanup is a guide, there are likely to be circumstances where containment in place of buried CWM is chosen over recovery and treatment of the buried CWM.

However, at sites where the initial decision is to leave CWM in place (with continued monitoring and institutional controls), the contents of buried CWM sites could be removed in the future if the existing site remedy is found not to be protective.

Direct Treatment Versus Storage of RCWM

RCWM differs from hazardous waste and other military munitions in one significant respect—the destruction of RCWM must be verified by the Chemical Weapons Convention (CWC) implementation body (just as the destruction of all CWM must be verified). The purpose of these verification provisions is to confirm, through on-site inspection, the accuracy of the relevant declarations; to ensure secure storage where storage is used; and to confirm that the CWM has been destroyed (Appendix IV (D)(37 and 50) of the CWC). Therefore, although storage may facilitate verification, there is nothing in the treaty or in the treaty implementation practice per se that requires storage as long as the CWC implementing body can adequately verify the destruction.[6]

Storage of RCWM from the large burial sites creates several challenges, including constructing interim holding facilities and providing the associated monitoring and security. In addition, regulatory requirements concerning storage times for hazardous waste may be difficult to comply with in the event that treatment and disposal operations are delayed due to technical or regulatory problems. More important, storage may involve increased risk.

---

[4]See generally National Contingency Plan, 40 CFR 300, 1990, as amended. The federal RCRA corrective action program uses essentially the same remedy selection criteria as required by Superfund. The hazardous waste cleanup programs of individual states often require the consideration of factors similar to those specified in Superfund. Some states have specified more stringent cleanup criteria and/or goals than the EPA (NRC, 1994).

[5]40 CFR 300.430(a)(1)(iii)(A-F)). Available online at <http://www.epa.gov/oilspill/pdfs/40cfr300.pdf>.

[6]Meeting between representatives of the Organisation for the Prohibition of Chemical Weapons and members of the committee at The Hague, The Netherlands, January 18, 2006.

According to the U.S. Army, there is probably sufficient flexibility in the CWC to devise adequate verification without necessitating storage, as long as such a verification plan is submitted to and approved by the CWC implementation group.[7] There is a precedent whereby a CWC oversight body permitted a nation to remove munitions from the ground and move it directly to destruction.[8] Of course, RCRA regulations require that hazardous waste must be adequately characterized prior to treatment, so some delay may be necessary while that is done. However, it would be logical for the U.S. Army, when possible, to use a process in which the munition is removed from the burial site, examined to determine whether it contains (or could contain) agent, and then destroyed immediately without intermediate storage. This process would minimize handling and eliminate the risk associated with storing potentially deteriorated munitions. The committee believes, therefore, that the U.S. Army could benefit from taking all possible steps to minimize the number of times munitions recovered from CWM burial sites are moved.

As a general rule, safety is maximized through minimizing the handling of deteriorated non-stockpile CWM. CWM recovered from the burial sites is likely to be in a deteriorated condition. This increases the likelihood of encountering leaking munitions and explosive instability due to the deterioration of explosive and propellant stabilizers, as well as potential contamination of explosives with toxic chemical agents or other contaminants. For these reasons, CWM items recovered from burial sites are likely to be more prone to leaking and explosive instability than CWM that has been stored under controlled conditions as part of the CWM stockpile. This makes it more hazardous to handle CWM recovered from burial sites than CWM from the stockpile program. The committee believes a serious accident (a major leak or accidental detonation) would be most likely to occur during the handling and transportation of deteriorated CWM contained in the large burial sites. Therefore, safety is maximized if the handling of the RCWM is minimized.

A remove-and-dispose approach will require that the CWM be removed from the large burial site no faster than it can be characterized and disposed of. The time it takes to characterize and dispose of the CWM will, therefore, determine the rate at which it can be removed from the large burial site. However, because many of the objects that will be removed from large burial sites will probably have been emptied of their contents prior to being placed in the burial pits, or because many may have leaked after placement in the pits and prior to their removal, they may not require much treatment before disposal.

This is a logical surmise because experience from other sites containing large amounts of non-stockpile CWM, such as Pine Bluff Arsenal, indicates that only between 10 percent and 25 percent of the CWM munitions and containers removed from the sites actually contain chemical agent.[9] This is likely to greatly reduce the need for CWM disposal for the objects removed from the large burial sites and would allow the removal of objects from the burial sites to proceed at a relatively rapid pace. Of course, the situation could change based on specific characterization data for the large burial sites.

PUBLIC INVOLVEMENT

One of the CERCLA and RCRA remedy selection factors noted earlier is community acceptance. It is, however, only a modifying factor. Community members can influence and sometimes take actions that lead to the rejection of remedial alternatives that they oppose, but they do not have the authority to veto otherwise scientifically acceptable alternatives.[10] There are also additional, more formal activities associated with federal facility restoration activities, such as funding of restoration site advisory boards composed of local citizens.

The activist public—consisting of local community organizations and national environmental coalitions—has played an important role in the selection of U.S. treatment technologies for CWM. Through the formal public involvement processes built into federal and state environmental statutes, litigation, and political action, citizens have influenced and delayed the deployment of technologies. Thus, an effective public involvement program is an essential component of any new strategy for the destruction of any type of CWM. The NSCMP's public involvement program has for several years served as a key element of mission success (NRC, 2004).

An effective public involvement approach has three components (NRC, 2002, 2004):

- Early and continuing provision of information to the public;
- Outreach, or opening channels of communication to allow the public to articulate its values, concerns, and needs; and
- Involvement, or providing mechanisms that engage members of the public and allow them to provide input and influence agency decisions.

The U.S. Army's NSCMP has implemented a national non-stockpile public outreach program, which includes distributing the NSCMP semiannual newsletter to interested

[7]Discussion between William R. Brankowitz, PMNSCMP, and the committee, Washington, D.C., November 29-30, 2005.

[8]Meeting between representatives of the Organisation for the Prohibition of Chemical Weapons and members of the committee at The Hague, The Netherlands, January 18, 2006.

[9]Conversation between William Brankowitz, NSCMP, and the committee on November 29, 2005.

[10]40 CFR 300.430 (f)(1)(i)(C). Available online at <http://www.epa.gov/oilspill/pdfs/40cfr300.pdf>.

parties throughout the nation, maintaining the NSCMP information repositories in 39 states and the District of Columbia, and maintaining the NSCMP Web site.[11] More importantly, it established the Non-Stockpile Core Group, a panel of community and environmental representatives plus state and federal officials. The Core Group, facilitated by the Keystone Center, provides regular comments to NSCMP on important components of the program (U.S. Army, 2004a).

In addition, NSCMP has worked with the U.S. Army Corps of Engineers and U.S. Army installations to conduct community meetings, such as occurred at Pine Bluff, Arkansas (NRC, 2002, 2004); to present briefings at restoration advisory board meetings, including at Spring Valley in Washington, D.C. (U.S. Army, 2003b); and to organize open house sessions, such as in Dover, Delaware (Merriweather, 2004).

Through such activities, the NSCMP has incorporated public concerns into its decisions and, in general, earned widespread activist support for its approach, particularly the use of mobile, nonincineration treatment systems such as the EDS and the RRS (CWWG, 2000). In previous reports (NRC, 2002, 2004), National Research Council committees found that the NSCMP's success in working with the public derives in large part from its willingness to consider the suggestions made by public stakeholders at both local and national venues.

FINDINGS AND RECOMMENDATIONS

The committee's review of buried CWM sites suggests several findings and recommended actions that may assist the Army in implementing the assessment and remediation of buried CWM.

Finding 2-1a. One of the factors to be considered in determining the appropriate remedy for an individual site is the short-term risk (e.g., the increased safety risk of handling accidents) posed by the remedial procedures.

Finding 2-1b. Eliminating or minimizing the storage of buried CWM that have been recovered will increase project safety and eliminate the potential technical problems caused by storage. That is, it will generally be preferable to characterize, remove, and immediately treat the RCWM and dispose of residuals in accordance with U.S. federal and state environmental requirements in a continuous process, rather than placing the RCWM in storage to await treatment and disposal at a later date.

Recommendation 2-1. When possible, the U.S. Army should adopt a continuous remove-and-dispose approach for the remediation of CWM from those large burial sites where the chosen remedy is removal and treatment, instead of removing them, storing them in an intermediate holding facility, and disposing of them later (the remove-store-dispose approach), as is being done for the non-stockpile CWM already recovered.

Finding 2-2a. The pace at which DOD, regulators, and/or Congress will require buried CWM to be recovered and destroyed is unknown. However, the rate of recovery and destruction of buried CWM, as determined by DOD, regulators, and/or Congress, will depend in part on the capabilities of existing U.S. technologies and the international technologies evaluated in this study.

Finding 2-2b. The U.S. Army's decision concerning what, if any, particular technology(ies) to use at a particular site depends on, among other things, the rate of recovery and destruction of buried CWM and the availability of technologies to safely, efficiently, and reliably treat potentially large numbers of recovered chemical munitions and chemically contaminated items.

Finding 2-2c. The U.S. Army's Non-Stockpile Program public involvement program (particularly the communication with the Core Group) and interactions with regulatory agencies have benefited the Army, the public, and the overall decision-making process. The Army's ability to use a new technology may depend on the public's view of the advantages and disadvantages of the candidate technology.

Recommendation 2-2. As the U.S. Army assesses and remediates buried CWM, it should continue its public involvement program, including use of the Core Group and interactions with regulators to solicit input on the (selection of) technologies that might be used for CWM destruction, particularly as a function of the rate of munitions recovery and destruction.

REFERENCES

CWWG (Chemical Weapons Working Group). 2000. Comments on the Draft Programmatic Environmental Impact Statement for Non-Stockpile Chemical Materiel Disposal. Available online at <http://www.cwwg.org/finaldpeis.html>. Last accessed April 24, 2006.

EPA (Environmental Protection Agency). 1997. Rules of Thumb for Superfund Remedy Selection, EPA 540/R/97-013/OSWER 9355.0-69, August. Available online at <http://www.epa.gov/superfund/resources/rules/rulesthm.pdf>. Last accessed February 14, 2006.

EPA. 2004. Treatment Technologies for Site Cleanup: Annual Status Report (11th Edition), EPA-542-R-03-009, February. Available online at <http://www.clu-in.org/download/remed/asr/11/asr.pdf>. Last accessed February 14, 2006.

Federal Register. 2003. Part IV, Department of Defense, 32 CFR Part 179, Munitions Response Site Prioritization Protocol; Proposed Rule. Federal Register 68(183): 50900-50941.

Federal Register. 2005. Munitions Response Site Prioritization Protocol—Final Rule. Federal Register 70(192): 58016-58054.

[11]For additional information, see <http://www.cma.army.mil/home.aspx>.

Merriweather, J. 2004. "How mustard gas shells are destroyed: Army to demolish ordnance found in Sussex driveway." DelawareOnline: The News Journal, October 7. Article ID wil2004100709090238.

NRC (National Research Council). 1994. Alternatives for Ground Water Cleanup. Washington, D.C: National Academy Press.

NRC. 2002. Systems and Technologies for the Treatment of Non-Stockpile Chemical Warfare Materiel. Washington, D.C.: National Academy Press.

NRC. 2004. Assessment of the Army Plan for the Pine Bluff Non-Stockpile Facility. Washington, D.C.: The National Academies Press.

U.S. Army. 2001. Final Programmatic Environmental Impact Statement, Transportable Treatment Systems for Non-Stockpile Chemical Warfare Materiel, February. Aberdeen Proving Ground, Md.: Product Manager for Non-Stockpile Chemical Materiel.

U.S. Army. 2003a. Army Environmental Cleanup Strategy, April. Available online at <http://www.asaie.army.mil/Public/ESOH/doc/Army-Environmental-Cleanup-Strategy-Final-28April03.pdf>. Last accessed April 24, 2006.

U.S. Army. 2003b. "June 2003 RAB meeting." Corps'pondent: A newsletter by the U.S. Army Corps of Engineers for the residents of Spring Valley. Available online at <http://www.nab.usace.army.mil/projects/WashingtonDC/springvalley/Newsletter/csjun-jul03.pdf>. Last accessed April 24, 2006.

U.S. Army. 2004a. Annual Status Report on the Disposal of Chemical Weapons and Material for Fiscal Year 2004. Available online at <http://www.pmacwa.army.mil/ip/dl/acwa_fy04_annual_report.pdf>. Last accessed February 13, 2006.

U.S. Army. 2004b. Recovered Chemical Warfare Materiel (RCWM) Response Process, EP 75-1-3, November 30. Available online at <http://www.usace.army.mil/usace-docs/eng-pamphlets/ep75-1-3/entire.pdf>. Last accessed February 14, 2006.

3

Evaluation Factors for International Destruction Technologies

The cultural, regulatory, and geopolitical environment in the United States often differs dramatically from that in other countries. Technologies that are acceptable in other countries may be unacceptable in the United States or may require upgrades or modifications as a condition of acceptance. Nowhere is this more evident than with technologies designed to destroy chemical weapons.

The complicated system of environmental, surety, and safety regulations prevailing in the United States also affects the acceptability of technologies that might be applied to the destruction of chemical weapons, as do the provisions of the Chemical Weapons Convention (CWC).[1] Although promising, a developing technology may not be a strong candidate in the United States if it cannot be tested, permitted, constructed, and operated in accordance with safety and environmental regulations and with the CWC. The bottom line is that acceptance of international technologies in the United States will depend on these technologies being mature, proven, efficient, reliable, robust, inherently safe, environmentally acceptable, and compatible with the provisions of the CWC, or holding great promise for being all of these things.

SELECTION OF EVALUATION FACTORS

In determining the evaluation factors to use for this study, the committee first examined factors used in similar evaluations in the past. The committee examined earlier NRC reports, including those pertaining to the U.S. Army's Alternative Technology Program and the Assembled Chemical Weapons Assessment Program.[2] Technology selection factors used in these reports address the special considerations associated with chemical weapons destruction and reflect concerns and issues raised by regulators and the public. After examining the system of comparative evaluation factors applied in these programs, the committee chose the factors used in one of these reports as a starting point for the present evaluation (NRC, 1995). The factors in that earlier report were modified for the present application to better reflect consideration of international technologies, non-stockpile issues, and present-day stakeholder concerns. Five primary factors were chosen for evaluation of the international technologies:

- Process maturity,
- Process efficacy/throughput,
- Process safety,
- Public and regulatory acceptability in a U.S. context, and
- Secondary waste issues.

A sixth factor, process costs, was considered, but in a more qualitative sense. Cost information was not generally available for many of the international technologies examined, and in any case, a complete and quantitative evaluation of cost was beyond the scope of the committee's task.

Each of the primary evaluation factors listed above has a number of subfactors, many of them at least somewhat interdependent. Each subfactor is expressed in the form of a question listed in a table for the evaluation factor to which it pertains, and its relationship to that evaluation factor is explained. For the Tier 1 technologies, the subfactor questions are then answered in five tables, one for each of the five evaluation factors.[3] The latter tables, which are contained in Appendixes B and C, provide some specific summary information and expert opinion in response to the questions. The information presented in this report, including that in Appendixes B and C, is based on the information that was

[1]See Chapter 2. Additional information on regulatory approval and permitting issues can be found in Chapter 4 of NRC (2002).

[2]The Assembled Chemical Weapons Assessment Program is now referred to as the Assembled Chemical Weapons Alternatives Program, although the acronym, ACWA, remains the same.

[3]Chapter 1 describes how the technologies were assigned to tiers.

TABLE 3-1 Process Maturity Subfactors

Subfactor	Relationship to Maturity
Is the technology in use for any type of material, even one not related to CWM in the U.S.?	If the technology is currently in use in the U.S. as described, the process is mature.
Is the technology in use for any type of material, even non-CWM related internationally?	If the technology is currently in use internationally as described, the process is mature.
Has the technology been permitted or otherwise approved in the U.S. for CWM or energetics?	If the process has been permitted or otherwise approved for treatment of CWM or energetic materials in the U.S., the technology is mature.
Has the technology been permitted or otherwise approved in the U.S. for industrial wastes?	If the process has been permitted or otherwise approved for treatment of industrial wastes in the U.S., it demonstrates that the technology is mature.
How much, if any, additional R&D is required in order to implement the technology?	If a moderate or an extensive amount of research and development is required to implement the technology, it may not be sufficiently mature.
What, if any, are the scale-up requirements needed to implement the technology?	Many technologies may be proven on a bench-scale or pilot plant basis, but significant scale-up issues may remain.
Can the technology be implemented within 3 to 5 years?	A technology should be capable of being permitted, constructed, and operated within a reasonable period of time.

available to the committee; some of the information was obtained from vendors and requires validation. The tables in Appendixes B and C allow a convenient side-by-side comparison of the various Tier 1 technologies with the respective Non-Stockpile Chemical Materiel Project (NSCMP) EDS or RRS/SCANS technologies currently in use. A more detailed discussion of the evaluation factors and associated subfactors is presented in the sections that follow here in Chapter 3.

DESCRIPTION OF EVALUATION FACTORS

Process Maturity

In general, chemical process technologies are located along a developmental continuum from laboratory-scale, proof-of-concept testing, pilot plant demonstration, and, ultimately, full-scale operation. Some technologies are in full-scale operation overseas at this time for the destruction of either CWM or industrial wastes. These or similar technologies may be in use in the United States, at least for industrial wastes. Still other technologies may not yet have reached this operational stage in the United States or elsewhere. Process maturity relates to whether the technology is being implemented in the United States or in other countries on a full-scale operational basis to deal with CWM or industrial wastes, and if it is not, to the nature and extent of the additional R&D that would be required to bring the technology to full-scale operation, specifically for non-stockpile materials. This maturity factor also considers whether technology implementation is feasible within a reasonable period of time. Process maturity subfactors are identified and their relationship to maturity is explained in Table 3-1.

Process Efficacy/Throughput

Process efficacy/throughput deals with the ability of the technology to destroy chemical agent and other process residuals (e.g., energetics) in a manner consistent with environmental regulations and CWC requirements. Destruction is typically evaluated by application of a concept initially established under the Resource Conservation and Recovery Act (RCRA) incineration regulations, known as destruction and removal efficiency (DRE) (40 CFR 264.343).[4] In general terms, DRE is the difference between the amount of chemical going into a process and the amount vented to the atmosphere after offgas treatment. The RCRA regulations for incineration require a DRE of 99.99 percent (otherwise known as four nines) for most organic constituents, but for some constituent categories, such as dioxins and furans, RCRA requires a DRE of 99.9999 percent (six nines). Incineration is capable of destroying chemical agents to a level of at least 99.9999 percent. Although incineration of chemical agents has lost favor with public interest groups in the United States, the incineration DRE of 99.9999 percent, as a practical matter, has become an informal basis of comparison for chemical agent destruction processes.

Process efficacy/throughput also involves process stability, reliability, and robustness. Considering the wide variety of non-stockpile materials that may be encountered in the future, and the conditions to which they may have been

[4]DRE for an incinerator is defined by the EPA as DRE = $[(W_{in}-W_{out})/W_{in}] \times 100\%$, where W_{in} = mass feed rate of a selected organic compound in the waste stream feeding the incinerator and W_{out} = mass emission rate of the same organic compound in the exhaust emissions prior to release to the atmosphere (40 CFR 264.343, July 1, 2004, edition).

TABLE 3-2 Process Efficacy/Throughput Subfactors

Subfactor	Relationship to Process Efficacy/Throughput
What is the DRE?	Technologies should be able to achieve a DRE for CWM of at least 99.9999 percent.
Does agent destruction meet the terms of the CWC (irreversible and verifiable)?	In accordance with the CWC, the physical, chemical, or biological reaction that destroys the CWM must be irreversible and verifiable.
What is the DRE for energetics?	Technologies should be able to meet a DRE for energetics of at least 99.99 percent.
Is the process reliable and robust?	The process should be able to operate with minimal downtime and should be reasonably insensitive to variations in process conditions. It should be able to complete operations in the event of a process upset (e.g., loss of power, mechanical problems).
Is the process highly complex or relatively simple?	It is often advantageous that a process be simple and easily explained to regulators and the lay public.
What are the personnel/staffing requirements for the technology?	The process should be able to be operated by personnel having a moderate level of education.
What is the process throughput?	Especially for large finds, the process should be able to treat a large number of munitions in a given amount of time.
Is the process scalable so that it can address small, medium, and large munition finds?	It is an advantage if models of various sizes and capacities can be applied to address finds of various sizes.
Is the process capable of handling multiple munition types?	It is an advantage if the process is capable of handling different types, sizes, and configurations of munitions in various states of disrepair and chemical decomposition.
Is the process capable of handling multiple agent types?	It is an advantage if the process is capable of handling different types of agents in various physical states.
Is the process transportable?	It is an advantage if the process equipment can be moved from site to site.

exposed for many years, this factor also addresses whether the technology is capable of handling a wide variety of munition types in various states of disrepair and chemical decomposition. Also, since it is often desirable in the United States to bring the process equipment to the munition (as opposed to transporting the munition to the process equipment), this factor pertains to process mobility. Finally, considering the potential need to remediate sites that may contain large quantities of buried non-stockpile munitions in the future, process throughput is also important. Process efficacy/throughput subfactors are identified and their relationship to process efficacy and throughput are explained in Table 3-2.

Process Safety

Process safety addresses specifically the ability of the technology, considering applied engineering controls and monitoring protocols, to ensure worker safety and also the safety of the surrounding community. Typically, evaluations of worker and community safety consider the risk of releases of chemical agent or process chemicals and the consequences of such releases. To more thoroughly evaluate such risks, maximum credible events are often postulated and the risks of such an event are assessed through quantitative risk assessments.

Some of the technologies entail operations performed either in tandem or in parallel. Such operations include the preparation of munitions for treatment; the storage and/or treatment of some secondary wastes, such as offgas; and equipment cleaning or maintenance operations. Quantitative risk assessments are sometimes used to ascertain and assign probabilities and consequences for accidents and other safety considerations for each phase of a total process—for example, munition accessing, treatment, and handling of secondary wastes. The committee had neither the time nor the resources to conduct such assessments but instead attempted to identify intrinsic safety issues associated with each technology and to qualitatively evaluate worker and community risk. Subfactors are identified in Table 3-3.

TABLE 3-3 Process Safety Subfactors

Subfactor	Relationship to Safety
What are the worker safety and health risks?	The process should be able to be operated with minimal risk to workers.
What are the community safety and health risks?	The process should be able to be operated with minimal risk to the community.
What are the process monitoring requirements?	Process monitoring should be no more complex than that for processes used in present U.S. non-stockpile applications.
To what extent have engineering controls been developed to ensure process safety?	Engineering controls should be sufficient to protect workers and the community from releases of chemical agent.

Public and Regulatory Acceptability in a U.S. Context

Earlier NRC reports on the non-stockpile program[5] identified regulatory approval and public involvement as key considerations for technology acceptance. Perhaps the most important consideration is that environmental regulators and the public should be involved in deciding whether to apply such technologies in the United States. Acceptability in a U.S. context also addresses specific concerns pertaining to chemical munitions destruction that have been raised by the U.S. public over the years, in both the stockpile and the non-stockpile programs.[6]

This factor also specifically evaluates environmental regulations established by the U.S. Environmental Protection Agency and the states for the destruction of chemical weapons and materials. It is critical that the technology be able to meet environmental permitting requirements and obtain environmental regulatory approval. There are a number of different regulatory approval and permitting (RAP) mechanisms that may be applicable to approving use of technologies to treat non-stockpile CWM. For example, technologies may be approved through different processes under the Comprehensive Environmental Response, Compensation, and Liability Act. Permits and other forms of regulatory approval may be issued under RCRA as well. The NRC has reviewed extensively RAP mechanisms that may be employed to approve technologies for treatment of non-stockpile chemical warfare materiel (NRC, 2002).

Regardless of which RAP mechanism is employed, the substantive permitting requirements of the RCRA program would need to be addressed in order for a technology to receive environmental regulatory approval. For example, if a technology were to be approved through the CERCLA remedial program, the substantive permitting requirements of the RCRA program would need to be addressed as an Applicable or Relevant and Appropriate Requirement, unless a waiver is obtained. The miscellaneous unit permitting requirements under RCRA[7] would likely apply since the international technologies under examination in this report would be unlikely to match any of the existing types of waste management units addressed under the RCRA regulations (40 CFR 264). The permitting requirements of the Clean Air Act, as well as the principles of pollution prevention and waste minimization, would apply as well. Acceptability subfactors are identified and described in Table 3-4.

Secondary Waste Issues

The term "secondary waste" encompasses a broad category of materials that are produced as a result of primary treatment. Technologies typically generate liquid wastes, various solids, and gaseous materials. Some of these materials can contain residual levels of chemical agent and other chemicals of concern, and additional treatment may be required. Such treatment may be conducted on-site (at the site of primary waste treatment), but commercial off-site treatment may also be considered. Storage and transportation requirements must also be considered. The generation of large volumes of secondary wastes contributes to adverse public reaction, and the analysis and certification that the wastes meet regulatory standards for disposal contributes to

[5]Review of the Army Non-Stockpile Chemical Materiel Disposal Program: Disposal of Chemical Agent Identification Sets (1999); Review and Evaluation of the Army Non-Stockpile Chemical Materiel Disposal Program: Disposal of Neutralent Wastes (2001); Evaluation of Alternative Technologies for Disposal of Liquid Wastes from the Explosive Destruction System (2001); Systems and Technologies for the Treatment of Non-Stockpile Chemical Warfare Material (2002); Assessment of the Army Plan for the Pine Bluff Non-Stockpile Facility (2004); Impact of Revised Airborne Exposure Limits on Non-Stockpile Chemical Materiel Program Activities (2005). All were published in Washington, D.C., by the National Academies Press.

[6]Public stakeholders are concerned, naturally, about things like process maturity, efficacy, and safety; however, the concerns raised under this evaluation factor have been raised specifically by public stakeholders in the United States in the past with regard to stockpile and non-stockpile operations.

[7]Since it is likely that technologies evaluated in this report will not be directly comparable to established technologies previously permitted under the RCRA program, technologies will need to meet the broad and stringent requirements pertaining to Miscellaneous Units established under 40 CFR Part 264, Subpart X.

TABLE 3-4 Public and Regulatory Acceptability in a U.S. Context Subfactors

Subfactor	Relationship to Public and Regulatory Acceptability in a U.S. Context
Is the process inherently incineration-like?	Some public stakeholders in the U.S. are opposed to use of incineration for the treatment of chemical warfare materiel.
Does the process break key chemical bonds (e.g., C-P bond for nerve agents)?	Regulators and other stakeholders in the U.S. have reacted favorably to technologies that result in complete destruction of key chemical bonds.
Could the process produce dioxins or other notable by-products?	Regulators and other stakeholders in the U.S. have reacted unfavorably to technologies that could create undesirable by-products.
Does the process allow holding and testing of process residuals prior to release?	Regulators and other stakeholders in the U.S. have reacted favorably to technologies that allow waste materials and by-products to be held and tested prior to their release.
Does the process result in excessive noise, odors, or other nuisances?	Regulators and other stakeholders in the U.S. have reacted unfavorably to technologies that are associated with excessive noise, odors, or other nuisances.
Would the process be able to satisfy environmental regulatory requirements under RCRA?	Permitting requirements under RCRA are stringent and have caused delays in technology implementation, particularly if there is public opposition (see NRC, 2002).
Would the process be able to satisfy environmental regulatory requirements under the Clean Air Act (CAA)?	Permitting requirements under the CAA are stringent and have caused excessive delays in technology implementation, particularly if there is public opposition (see NRC, 2002).
Would the process be able to satisfy other applicable environmental regulatory requirements?	Some technologies may require compliance with other environmental laws such as the Clean Water Act.
Does the process satisfy the principals of pollution prevention and waste minimization?	Technologies, to the extent possible, should employ process chemicals that are nontoxic, and the technology should result in minimal amounts of secondary wastes.

processing costs. In addition, residuals from secondary waste treatment may require further treatment prior to disposal. Secondary waste issues have taken on great importance, especially over the last few years. A prime example of this is the concern that has been in evidence over the Army's plans to send hydrolysate resulting from treatment of bulk VX from the Newport Chemical Depot to a facility located in New Jersey along the Delaware River (Ember, 2005). Subfactors for secondary waste issues are identified and described in Table 3-5.

Process Costs

Technology costs can be evaluated on a number of different levels and from a number of different perspectives. For example, costs can be evaluated on a per-munition basis, on a per-site basis, or as a function of the amount of chemical agent. The basis for the cost evaluation is often an important element in the evaluation of cost. For example, as the size and complexity of the chemical munitions removal and cleanup increases, permanent or semipermanent treatment facilities, which are more expensive and complicated than mobile treatment units, become more reasonable. Although cost is a consideration in technology selection, it is important to note that a relatively high cost does not necessarily mean that a technology is unacceptable, especially if no technically and socially acceptable alternative is available.

Realistic cost information can be difficult to obtain, particularly for technologies that have not yet achieved production scale and for which operating experience is limited. As previously indicated, cost information was not generally available to the committee for the international technologies evaluated in this report, so it was not possible for the committee to conduct life-cycle cost analyses for these technologies. Consequently, the committee chose to evaluate costs only in a qualitative sense, with a focus on identifying those cost components of a system or technology that might be associated with a relatively high cost. An example would be the potentially high energy costs of some technologies. A summary paragraph discussing these types of considerations is presented toward the end of the Tier 1 technology evaluations. The committee assumes that if the U.S. Army chose to further consider an international technology for implementation in the U.S. non-stockpile program, it would require detailed cost estimates before proceeding with further technology research or implementation.

RATING SYSTEM

The committee determined that it would be useful to develop a rating system to enable efficient comparative evaluation of the technologies with respect to each of the evaluation factors and, ultimately, of the technologies themselves and the current NSCMP equipment in use. Because

TABLE 3-5 Secondary Waste Issues Subfactors

Subfactor	Relationship to Secondary Waste Issues
What is the character of secondary wastes? Form (e.g., liquid, solid, gas) Volume Toxicity (e.g., agent, degradation products, metals, other contaminants)	Secondary waste issues are most significant for wastes generated in large volume or that may contain residual amounts of agent, agent degradation products that retain some toxicity, and other contaminants of concern.
Do secondary wastes initially meet: General population limits (GPLs) or short-term exposure limits (STELs)? CWC requirements? Environmental regulatory requirements?	Secondary wastes that are generated, and in particular solids, must meet the Army's requirements for decontamination. Wastes that meet GPLs might be treated as non-hazardous wastes or recycled without further controls. Secondary wastes that meet STELs but not GPLs require additional management.[a] Secondary wastes may need additional scrutiny under the CWC if they contain Schedule 2 chemicals.[b] Additional treatment may be required if secondary wastes do not meet environmental regulatory requirements as generated.[c]
For each secondary waste, will subsequent treatment be required: To meet GPLs or STELs? To satisfy CWC requirements? To satisfy environmental regulatory requirements?	If additional treatment is required to meet the various listed requirements, such treatment presents additional risk and costs. The wastes may need to be transported to the site of treatment, and additional storage may be required.
For each secondary waste, if subsequent treatment is needed, are treatment methods established and available?	If secondary wastes require additional treatment, acceptable means of treating these wastes must be available.
Will residuals from treatment of secondary waste require subsequent treatment: To meet GPLs or STELs? To satisfy CWC requirements? To satisfy environmental regulatory requirements?	In some cases, even residuals from secondary waste treatment may require additional treatment to meet the various standards listed.
What is the disposition of final treatment residuals: Recycle? Hazardous waste landfill? Nonhazardous waste landfill? Other?	Some secondary wastes, even after treatment, may be considered hazardous and may need to be disposed of accordingly. Some types of secondary wastes may be released as is for reuse or recycling.

[a]General population limits (GPLs) and short-term exposure limits (STELs) are collectively termed airborne exposure limits (AELs) and are used by the Army as a means of protecting workers, the general public, and emergency responders from the toxic effects of airborne exposure to chemical agents. Application of AELs was reviewed extensively in NRC (2005).

[b]The CWC established a schedule of chemicals that are controlled under the CWC. Several of the agent degradation products are designated under CWC Schedule 2, and their manufacture and distribution in commerce is controlled. If secondary wastes contain Schedule 2 chemicals, additional scrutiny from CWC inspectors may be required during secondary waste treatment or disposal.

[c]Some secondary wastes may contain hazardous waste constituents (e.g., heavy metals) regulated under the RCRA program, and if such contaminants are present above certain concentrations, may require additional treatment prior to ultimate disposal.

the technologies differed in terms of data and information available to the committee, and considering that some of the technologies were in different stages of development and/or implementation, the committee developed the following qualitative rating system:

- \+ Fully acceptable. Indicates that no or only minor issues remain with respect to any one evaluation factor or the technology as a whole.
- 0 Partially acceptable. Indicates that some issues remain with respect to any one evaluation factor or with a technology as a whole but that, in general, these issues should be resolvable.
- – Unacceptable. Indicates that some issues remain with respect to any one evaluation factor or the technology as a whole, and that these issues are unlikely to be resolved favorably.
- ? Inadequate information. Indicates that not enough information was available to fully evaluate the technology with respect to any one evaluation factor or the technology as a whole. This rating may also indicate that information was available but was classified, proprietary, or otherwise restricted from public dissemination.

TABLE 3-6 Statement of Task Directives and Corresponding Technology Evaluation Factors

SOT Directive	Evaluation Factor
Potential to be more effective for the overall disposal of specific types of non-stockpile materiel	Process efficacy/throughput
Assessment of technical feasibility	Process maturity and process efficacy/throughput
Assessment of level of maturity	Process maturity
Assessment of degree of scientific acceptance	Process maturity and process efficacy/throughput
Implementation and deployment issues related to cost	Process costs (qualitative analysis only)
Implementation and deployment issues related to safety	Process safety
Implementation and deployment issues related to risk	Process safety
Implementation and deployment issues related to protection of the environment	Public and regulator acceptability in a U.S. context and secondary waste issues
Acceptability to regulators and stakeholders	Public and regulator acceptability in a U.S. context and secondary waste issues

The committee recognizes that this rating system, and any similar system, is necessarily subjective. In addition, because several of the technology providers did not have or could not give out certain information to the committee (owing to proprietary considerations, for example), the ratings may not fully represent the acceptability of the technology with respect to any one factor, or as a whole. Before decisions are made about any technology, such as whether or not to further consider use its in the United States, the committee would urge a more in-depth evaluation, especially taking into consideration information that was restricted from public dissemination.

ASSESSMENT OF EVALUATION FACTORS AGAINST DIRECTIVES REFLECTED IN THE STATEMENT OF TASK

The committee believes that the overall system of factors and subfactors used in this report encompasses the directives reflected in the statement of task. Table 3-6 identifies directives from the statement of task and shows which of the evaluation factors specifically address those directives.

REFERENCES

Ember, L.R. 2005. "Army halts VX destruction." Chemical & Engineering News 83(28): 13.

NRC (National Research Council). 1995. Evaluation of the Army's Draft Assessment Criteria to Aid in the Selection of Alternative Technologies for Chemical Demilitarization. Washington, D.C.: National Academy Press.

NRC. 2002. Systems and Technologies for the Treatment of Non-Stockpile Chemical Warfare Materiel. Washington, D.C.: National Academy Press.

NRC. 2005. Impact of Revised Airborne Exposure Limits on Non-Stockpile Chemical Materiel Program Activities. Washington, D.C.: The National Academies Press.

4

Tier 1 International Munitions Processing Technologies

INTRODUCTION

In the course of its information gathering, the committee recognized that one particular type of international technology has risen to prominence in addressing the cleanup of old and abandoned chemical weapons at sites in other countries. Detonation-type destruction technologies rely on the ability of the energy from explosive charges within a containment vessel to efficiently destroy recovered chemical munitions and the agent and energetics contained therein.

There are several versions of detonation-type technologies. An earlier version of the controlled detonation chamber (CDC) was reviewed by a previous National Research Council committee.[1] Since then, this technology has undergone further development and implementation in several European venues. Meanwhile, two more recent examples of detonation-type technologies that are in use or being developed for destroying recovered chemical warfare munitions have come to the committee's attention, namely, the Japanese detonation of ammunition in vacuum integrated chamber (DAVINCH) technology and the Swedish Dynasafe technology. The committee considers these two technologies and the latest CDC technology as sufficiently capable and mature to warrant Tier 1 status for further consideration by the Non-Stockpile Chemical Materiel Project (NSCMP) as an alternative to the explosive destruction system (EDS) currently used by NSCMP, or as a complementary means of processing recovered non-stockpile munitions.

MEASUREMENT OF PERFORMANCE FOR DETONATION TECHNOLOGIES

A discussion of the Tier 1 detonation-type technologies will be informed by first considering appropriate means for gauging their performance. A measure of performance for detonation processes would be useful to the U.S. Army because it would allow comparing the relative effectiveness of different technologies. Also, although the level of performance and the precise test used to measure such performance is ultimately a decision for federal and state regulators, any information the U.S. Army might obtain or generate on the performance of these technologies would certainly be helpful in obtaining regulatory approvals to deploy such technologies. Moreover, the process of developing a detailed test procedure could form the basis for reaching a consensus with regulators. Furthermore, many members of the public interested in the destruction of CWM distinguish between destruction efficiency (DE) and destruction and removal efficiency (DRE).[2] Thus, an accepted measure of performance for detonation technologies will assist the Army in addressing questions from the public (see also discussion of public involvement in Chapter 2 and DREs in Chapter 3).

However, determining such a measure of performance for detonation processes appears to offer unusual challenges, and, based on the information available to the committee, the committee believes the Army should specify requisite documentation from vendors and employ engineering contractors to review it to determine if the data provide a consistent and reliable measure of performance. For other processes,

[1]See the National Research Council report *Systems and Technologies for the Treatment of Non-Stockpile Chemical Warfare Materiel* (2002).

[2]For a definition of destruction efficiency, see <http://www.basel.int/techmatters/popguid_may2004_uk_pros%20and%20cons.pdf>.

$$DE = 100 \times ((\text{input} - \text{output})/\text{input})$$

For destruction of a chemical weapon, input would be the quantity of agent in a munition and output would be the quantity of agent in all the final residual streams after the detonation process has destroyed that munition. For comparison, the definition of destruction and removal efficiency is

$$DRE = 100 \times [(\text{feed rate} - \text{emission rate})/(\text{feed rate})]$$

where emission rate is the rate at which the selected organic compound exits the process in the exhaust gas stream. The DRE thus focuses on air emissions while DE focuses on total destruction.

procedures have been established or are obvious and straightforward. Thus, the trial burn approach is well established for incinerators. A selected organic compound (which is more difficult to destroy than the typical waste burned in the incinerator during normal permitted operation) is fed at a known rate to the process. The mass of each effluent stream is measured, along with the concentration of the selected organic compound. The degree of destruction is then calculated. For incinerators, this is the DRE, which refers to "the percent of waste material that is either destroyed or otherwise removed from the waste feed" (ATSDR, 2005, p. 18).

In the equation DRE = 100 × [(feed rate − emission rate)/(feed rate)], the feed rate is the measured amount of chemical in the wastes fed to the incinerator and the emission rate is the measured amount of a chemical in the stack exhaust (ATSDR, 2005). The DRE measures the effectiveness of the treatment process as a whole.

For neutralization, hydrolysis, and many other processes that treat agent, the procedure is straightforward. Agent is fed at a known rate or in a known amount to the process. The mass of each effluent stream is measured, along with the concentration of the agent. Generally, there is no formal DRE that applies to neutralization and hydrolysis processes, although one can perform such a calculation.

Detonation processes destroy whole munitions, in discrete events. A procedure for determining the degree of destruction for a detonation process should ideally involve feeding complete munitions into the process; the feeding of neat agent in place of complete munitions would not give meaningful information.[3]

One possible approach involves determining the mass of the liquid in the munitions and the concentration of agent in the liquid, then measuring the mass and agent concentration in all the streams leaving the process. This approach could also involve measuring agent retained within the system, i.e., within the detonation chamber, but this could be difficult. Information thus obtained could then be used to calculate the DRE. The committee anticipates that the DRE will be a more important number than the DE. It would also be helpful to gather and report additional information gained from analysis of effluent streams, such as quantity of dioxins and furans produced, quantities of Schedule 2 compounds, and the proportions of the three valence states of arsenic. Comparison of these measurements with similar EDS performance measurements would also be important.

The DRE reflects how well the offgas management system is designed as well as how effectively the detonation destroys agent. Both are important. In evaluating detonation-type technologies, the degree of agent destruction in the actual detonation event should be measured. Of course, permits and regulatory approvals of such systems will typically entail process monitoring to ensure that they are operating as designed. Hence, in addition to being able to demonstrate an acceptable DRE, technologies must be able to demonstrate that agent is effectively destroyed and that secondary waste streams, including gases vented into the atmosphere, do not contain agent above agreed-on levels.

CONTROLLED DETONATION CHAMBER TECHNOLOGY

Description

The CDC, previously known as the Donovan blast chamber or the contained detonation chamber, was developed and is manufactured by DeMil International, Inc., of Huntsville, Alabama. The CDC was applied earlier to replace open detonation operations for destruction of conventional high-explosive munitions. It provides a contained environment that prevents the release of blast fragments, heavy metals, and energetic by-products. It was later proposed that a CDC could be used to destroy chemical warfare materiel (CWM) by detonation in its enclosed environment. The working assumption was that the heat and pressure of a contained explosion would destroy the chemical agent, especially in the wet environment produced by inclusion of water bags in the detonation chamber. Initial tests on World War I munitions recovered in Belgium indicated that a high level of agent destruction could be achieved. The preliminary results were reviewed in an NRC report (NRC, 2002).

Following the encouraging results of the Belgian tests, the U.S. Army has supported further testing in cooperation with the British Defence Science and Technology Laboratory at Porton Down, England. This further testing involved extensive modification of the basic Donovan blast chamber system to make it suitable for destruction of chemical munitions in an U.S. regulatory context. The Belgian tests were performed with a relatively small T-10 unit that had undergone only modest modifications to make it suitable for destroying toxic chemicals. The systems that have evolved from the Porton Down tests are much larger (requiring two 40-foot trailers for transport of the TC-25 or eight for the TC-60 vs. one for the T-10). The larger systems can process larger weapons, and most of the manual handling of munitions has been eliminated (Bixler, 2005).

Description of Original Test Unit

As tested in Belgium, the CDC consisted of three main components: the detonation chamber, an expansion chamber, and an emissions control unit, the latter comprising a particle filter and a bank of activated carbon adsorption beds (NRC, 2002). The maximum explosive rating of the T-10 mobile unit is 12 pounds of TNT-equivalent, including the donor charge used to access the burster and the agent.

The detonation chamber is connected to a larger expansion chamber. A projectile wrapped in explosive is mounted in the

[3]As used here "complete munitions" means munitions containing either agent or a chemical surrogate that is more difficult to destroy than the chemical agent that is most resistant to destruction.

detonation chamber. The floor of the chamber is covered with pea gravel, which absorbs some of the blast energy. The gravel is renewed periodically because it fractures during the explosions. Bags containing water are suspended near the projectile to help absorb blast energy and to produce steam, which reacts with agent vapors. After the detonation chamber is loaded, its entry port is sealed and the exit from the expansion chamber is closed. After the explosive is detonated, the chambers are kept sealed for about 2 minutes to maintain heat and pressure. The gases are then vented through the main duct to the baghouse and the carbon adsorption beds. Gases are monitored at several points in the CDC system for agent, carbon monoxide, and volatile organics as well as for agent at the exit duct outlet. The concentrations of particulates suspended in the vapors, such as soot, gravel dust, and metal oxides, were also monitored during the Phase 1 tests (De Bisschop and Blades, 2002). Water vapor from the explosives and from the explosion-quenching water bags collects on the charcoal filters.[4]

After the detonation, the atmosphere in the detonation chamber clears fairly rapidly as air is drawn through the system to remove residual organic vapors, thereby permitting reentry for placement of the next round. During the tests in Belgium, 15 chemical munitions were treated in the CDC in 3 hours, including 20-minute breaks after every five munitions (U.S. Army, 2001). This amounted to an average treatment time of 12 minutes per munition, including the time for breaks. Analysis of the pea gravel and of wipe samples from the chamber walls showed low agent concentrations (1.2 to 64.4 mg/kg in pea gravel; 0.39 to 78.65 mg/m^2 in wipe samples from detonation chamber) during the Belgian test series (De Bisschop and Blades, 2002).

The main waste materials from destroying chemical munitions were solids: soot, charcoal (from the filters), pea gravel, inorganic dust, and metal fragments from the weapons. The major liquid waste from the CDC was spent hypochlorite solution from decontamination of the system prior to maintenance operations.[5] The solids, which may have been contaminated with traces of chemical agent and explosives residues, were packaged in plastic bags and placed in shipping containers that were sent to a commercial hazardous waste incinerator for disposal.

Current TC-25 and TC-60 Chemical Munitions Destruction Units

The CDC T-10 model tested in Belgium can treat complete chemical munitions up to 105-mm in diameter. A larger mobile unit (TC-25) was tested extensively at Porton Down, England (Blades et al., 2004) (see Figure 4-1). A still larger unit (TC-60) with an explosive capacity of 60 pounds of TNT-equivalent is now available (Bixler, 2005). It can handle munitions over 200 mm in diameter, according to the manufacturer. Table 4-1 provides the dimensions of the pressure chambers for the three CDC models.

The latest versions incorporate a mechanical system to move explosive-encased munitions from the preparation area through a reduced pressure vestibule into the detonation chamber. Double doors on the detonation chamber minimize any chance that agent vapors or detonation debris might escape. For standard varieties of munitions, the explosive charge is precast in a plastic form that can be slipped over the projectile. This packaging mode minimizes worker contact with the munitions and facilitates the mechanical transport of the projectile into the detonation chamber. Nonstandard items may require wrapping the munitions in sheet explosive, as was done in Belgium.

In the detonation chamber itself, armor plate can be affixed to the walls to reduce the likelihood of damage by flying metal fragments. The experience to date suggests that the chamber will retain full integrity for thousands of shots. Predicted lifetime is greater than 200,000 shots (Bixler, 2005). Injection of hot air or gaseous oxygen into the detonation and expansion chambers facilitates decomposition of any chemical agent adhering to the walls or adsorbed on the pea gravel or other solids.

A significant change in operating procedure from that used in the Belgian tests is applied in decontaminating the chambers in preparation for maintenance. In the early tests, the walls of the chambers and the pea gravel were washed with sodium hypochlorite (bleach) solution to oxidize any residual chemical agent. This procedure was effective but required much manual effort and resulted in a liquid waste that required separate disposal. In the revised procedure, the chambers are flushed with hot (450°F) air for up to 24 hours to destroy residual agent. An alternative procedure is to detonate a small explosive charge that destroys the residual agent thermally. Both procedures reduce worker exposure and eliminate the generation of a liquid waste stream (Bixler, 2005).

The back end of the system, into which the offgases from the expansion chamber vent, has also been modified extensively (Blades et al., 2004). The vapors and particulates arising from the detonation of the munition pass through a reactive-bed filter (hydrated lime or sodium bicarbonate) to remove acidic gases and a porous ceramic filter to collect particulates, including soot and dust from the pea gravel. A lime precoating on the ceramic scavenges acidic vapors

---

[4]The committee noted that water vapor competes with organic species for sites on the charcoal filters. Saturation of these sites with water vapor could reduce the effectiveness of the filters in removing organic species from the emission stream (NRC, 2002). In the current system, agent monitoring between the two series-mounted carbon filter beds can detect overloading of the first filter bed before any possible breakthrough from the overall system.

[5]Personal communications between Herbert C. De Bisschop, Belgian Military Academy, and George W. Parshall, July 25, 2001.

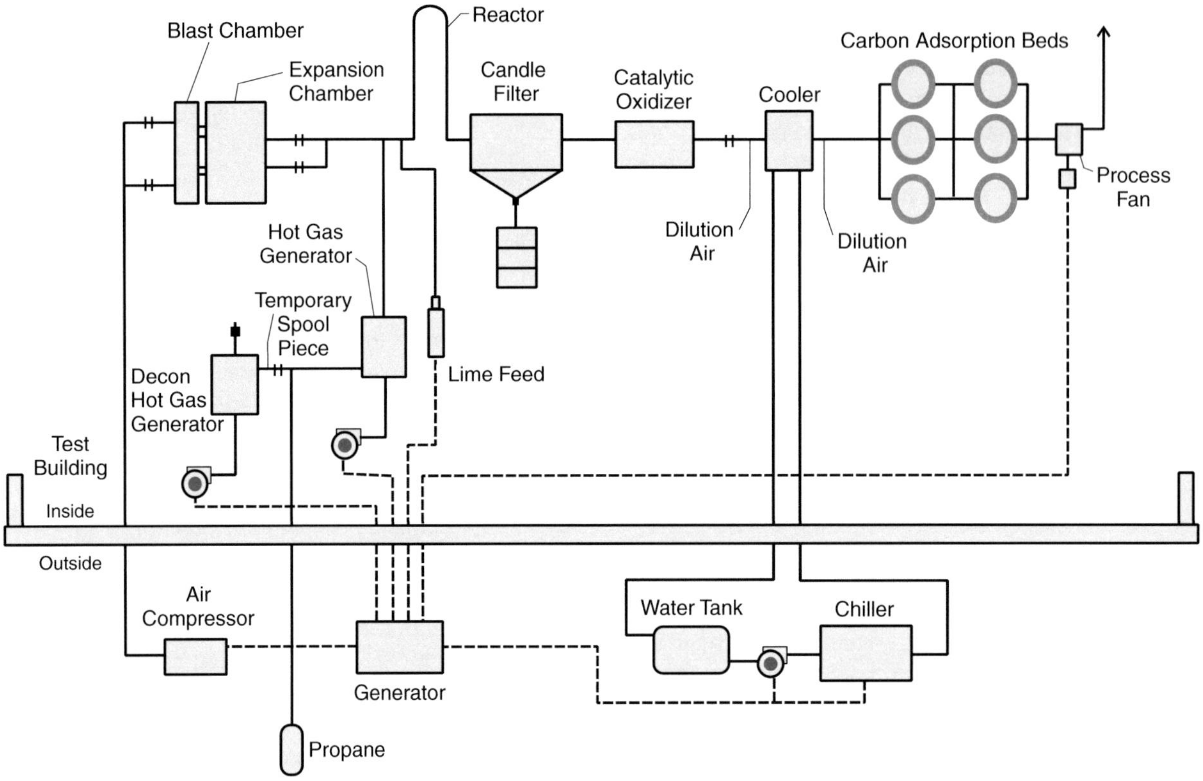

FIGURE 4-1 TC-25 CDC system layout. SOURCE: Blades et al., 2004.

TABLE 4-1 Dimensions of the Pressure Chambers in Three CDC Models Designed for Destroying Chemical Warfare Agents

	Detonation Chamber		Expansion Tank		
CDC Model	Interior (m)	Volume (m^3)	Interior (m)	Volume (m^3)	Total Volume (m^3)
T-10	1.524 × 1.524 × 1.524	3.5	2 × 2 × 2.3	9.2	12.7
TC-25	1.981 × 2.286 × 2.845	12.9	2.438 × 2.438 × 10.515	62.5	75.4
TC-60 PD	2.438 × 2.438 × 3.657	21.5	2.286 dia × 10.516[a]	43.1	64.6

[a]The expansion tank for Model TC-60 PD is cylindrical.

SOURCE: Briefing by CH2MHILL to Thales and the Délegation Générale pour l'Armament, October 2005.

that escape the reactive filter. A catalytic oxidation unit (CATOX)[6] oxidizes carbon monoxide and organic vapors from the gas stream prior to venting through a two-stage carbon adsorption bed system. MINICAMS[7] monitoring of the gas stream indicates that no detectable agent reaches the adsorption bed.[8]

[6]A CATOX unit facilitates the oxidation of carbon monoxide, hydrogen, and volatile organic compounds contained in an air stream such as that emerging from the particle filter in the pollution control system of the CDC. Generally, the air stream is passed through a bed of a catalytic solid that acts very much like that in an automotive catalytic converter.

[7]A MINICAMS is an automatic, near-real-time continuous air monitoring system using gas chromatography and sample collection with a solid-adsorbent preconcentrator or fixed-volume sample loop. The MINICAMS collects an air sample, performs an analysis, and reports the result. Reported agent concentrations above a user-set threshold generate an alarm status, which can be reported in various ways (see <http://www.oico.com/default.aspx?id=product&productID=75>).

[8]Controlled detonation chamber (CDC) update. Briefing by DeMil International to the Non-Stockpile Program Core Users Group, November 2004.

Country-by-Country Experience

Belgium is the only nation in which the CDC has been used in a production mode for destroying chemical weapons. Although tests were carried out with a variety of World War I chemical agents and munitions, the CDC has been used primarily to destroy German 77-mm artillery projectiles containing Clark II (diphenylcyanoarsine) agent, an arsenical irritant. The system has been generally satisfactory, and over 2,000 such projectiles have been destroyed in 5 years.

The United States and the United Kingdom have collaborated on a series of tests that demonstrated the ability of a transportable CDC to safely destroy other chemical munitions that may be found at sites in the United States and the United Kingdom (Blades et al., 2004). Many improvements have been made to the CDC system to reduce manual operations, to simplify waste disposal, and to ensure that chemical agent vapors do not escape into the environment. Pending successful completion of a test series under way in early 2006, the system should be ready for implementation if it proves cost effective and publicly acceptable.

Evaluation Factors Analysis for CDC

Process Maturity

The use of the CDC to destroy chemical munitions has been demonstrated in a series of campaigns over a 5-year period. As mentioned above, the first tests were carried out in Belgium in May and June 2001. During those tests, live munitions containing sulfur mustard agent, Clark arsenical agent, and phosgene were destroyed. The original Donovan CDC system and the operating procedure were modified to enhance worker safety and reduce potential emissions of residual chemical agent or agent decomposition products. Extensive monitoring was conducted to determine agent DE and establish the quantity and nature of the decomposition products (De Bisschop and Blades, 2002).

Subsequently, the Belgian military used the TC-60 CDC in a production mode to destroy part of its large stockpile of recovered chemical warfare materiel (RCWM) at Poelkapelle. Over 2,000 German 77-mm projectiles containing Clark arsenical agents were destroyed in the T-60 unit (Bixler, 2005).

Following the success of the Belgian testing, the U.S. Army supported a series of tests at Porton Down in the United Kingdom to demonstrate the usefulness of the CDC for operations in the United States. These tests included modifications of the system to enhance DE, to improve worker safety, to improve productivity, and to minimize any possibility for escape of agent vapors.

Phase I testing was carried out from April to September 2003 (Blades et al., 2004). A variety of munition types containing sulfur mustard agent, phosgene, a phosgene-chloropicrin mixture, and a smoke composition were destroyed.

Phase II demonstration/validation testing was conducted at Porton Down in 2004 (Bixler, 2005). The tests included detonation of two munitions per shot, a key point in establishing the potential throughput of the CDC. Extensive computer control and safety interlocks were added to regulate contact of any agent vapors with the treatment system and to remove any opportunity for a detonation to occur before the complete system is ready for operations.

Another series of tests at Porton Down was scheduled for early 2006. A major goal of these demonstrations was to demonstrate the potential throughput of the TC-60 CDC. Modeling indicates that 22 shots (up to 40 munitions)[9] can be conducted in a 10-hour shift (DeMil International, 2005a).

Process Efficacy/Throughput

The CDC appears to be well suited for destroying a range of either chemical or conventional munitions (NRC, 2002). While it has yet to be tested for the destruction of nerve agents (cf. Table B-2), the hot, wet, oxidizing atmosphere in its detonation chamber can reasonably be expected to decompose these compounds rapidly. The CDC has also not been demonstrated for munitions encased in overpacks for storage.

The DE achieved by the detonation alone appeared to be above 99 percent, as measured by the postdetonation environment in the Belgian tests (De Bisschop and Blades, 2002). A similar analysis done in the U.S. Army/U.K. Defence Science and Technology Laboratory tests gave a DE from detonation of 99.408 to 99.998 percent in a series of five tests with HD-loaded 4.2-inch mortars. In five tests in which agent destruction was enhanced by the addition of gaseous oxygen to the detonation chamber prior to the blast, the DEs from detonation ranged from 99.965 to 99.996 percent.[10] These calculated efficiencies were based on measurement of residual agent in the pea gravel and the walls of the detonation chamber. No residual agent was found downstream in the expansion chamber or the pollution control system.

The more important measure from the viewpoint of preventing releases that might endanger workers, the public, or the environment is the DRE. No published DRE figure has been found, but it is likely to be as least 99.9999 percent ("six nines") because the posttreatments reduce agent concentrations to below detectable levels as measured by a MINICAMS before the offgases reach the carbon adsorption beds (Bixler, 2005).[11] It does not, however, qualify as a hold-and-test system like the EDS.

---

[9]Multiple 75-mm projectiles or 4.2-in. mortars can be treated in a single detonation operation.

[10]Brint Bixler, CH2MHILL, responses to committee questions of February 6, 2006.

[11]Although the reference does not provide a method detection limit for the MINICAMS as used in this situation, the MINICAMS can generally detect HD at levels of 0.001 mg/m^3 and sometimes lower (NRC, 2005).

TABLE 4-2 Estimated Throughput Rates for CDC TC-60

Munition	Munitions per Cycle	Cycles per 10-hr Day	Munitions per 10-hr Day
4.2-in. mortar, M1	2	20	40
75-mm projectile, M64	2	20	40
5-in. projectile, MK VI	1	22	22
5-in. projectile, MK 54	1	22	22
155-mm projectile, MK II	1	22	22
8-in. projectile, T174	1	22	22
Bomblet, M139	3	20	60
105-mm projectile, M60	1	22	22
100-lb bomb, M47	—	30	6[a]
115-lb bomb, M70	—	30	5[b]

[a]Agent drained into five 20-lb lots; each lot detonated in CDC. Five 20-lb lots/bomb × 6 bombs/day = 30 cycles/day.
[b]Agent drained into six 20-lb lots; each lot detonated in CDC. Six 20-lb lots/bomb × 5 bombs/day = 30 cycles/day.

SOURCE: CH2MHILL, responses to committee questions of February 6, 2006.

Models of the CDC up to the TC-60 are designed to be transportable although there may be some restrictions on road transport because of the physical size of the detonation chamber. These models are designed to be set up within 5 days. The typical operating crew comprises 18 staff, including laboratory, safety and supervisory personnel (DeMil International, 2005b).

Because there is no time-consuming neutralization step, the CDC's throughput could be much higher than that of the EDS, which conducts only one detonation every other day. However, the comparison is complicated by the fact that the EDS can destroy more than one munition per shot, depending on the size of the munitions. The EDS-1 can handle three mortar rounds, and the EDS-2 has destroyed as many as six per shot. As noted above, the CDC has demonstrated destruction of two munitions per shot and could potentially destroy 40 projectiles per 10-hour shift. Estimated throughput rates per 10-hour day for representative U.S. munitions are shown in Table 4-2. The current CDC also has the advantage in operation of generating little or no liquid waste that requires subsequent processing, in contrast with the substantial neutralent and rinsate effluents produced with the EDS.

Process Safety

The continuing development of the CDC has significantly reduced the manual operations in the treatment of CWM. The original T-10 system tested in Belgium involved personal protective equipment (PPE)-clad workers in operations such as wrapping projectiles in sheet explosive, moving the projectile into the detonation chamber, and connecting fuzes and detonators. After detonation and cooling of the chamber, the workers had to prepare the chamber for reloading despite the presence of traces of agent on the chamber walls and the pea gravel. Preparation for weekly maintenance operations included washing the walls and floor of the chamber with decontamination solution. Workers also packed agent-contaminated filter material for shipment to a TSDF (De Bisschop and Blades, 2002).

The modifications applied during the Porton Down tests reduced manual operations by slipping precast donor explosives over the projectile and mechanically moving the round into the detonation chamber. Even in the advanced TC-60 system, however, there remains a manual step. Between shots, an operator must reach inside the door to the detonation chamber to unplug the electrical connector for the detonator from the last detonation, then plug in the connector for the next detonation. This approach might slightly increase the potential for worker exposure, but it eliminates the chance of mechanical failure of an automated plug connection system.

Routine munition preparation operations are conducted by workers in Level C PPE. Level B PPE, offering a higher level of protection than Level C, is used for maintenance work in and around the chambers (Blades et al., 2004). A process hazards analysis for the current TC-60 model was conducted in mid-2005 (DeMil International, 2004). According to the technology proponent, it was a "qualitative analysis prepared in accordance with U.S. Army's AR 385-64 and AR 385-61 directives, and Guidelines for Hazard Evaluation Procedures. . . ."[12] The analysis covered an extensive range of operations, failure modes, and corrective actions and provided qualitative severity assessments of failure modes. Supporting systems such as that which supplies oxygen to the detonation chamber were included in the evaluations and process modifications. It was reported by the technology vendor that this process hazards analysis had been reviewed

[12]Brint Bixler, CH2MHILL, responses to committee questions of February 6, 2006.

and agreed with by the U.S. Army's Edgewood Chemical Biological Center.[13]

The substitution of hot air purging for washing the chamber and detonation debris with decontamination solution removed a set of operations that probably constituted a significant risk of agent exposure. The improvements to the pollution control system seem to have minimized agent-contaminated waste materials (Bixler, 2005).

Public and Regulatory Acceptability in a U.S. Context

The CDC has not been permitted for use in destroying CWM in the United States, although it has been used successfully in Europe. Additional testing of the CDC may be required if the system is to be permitted in the United States for treatment of CWM. The system's DE from detonation of 99 to 99.99 percent is modest; the DRE of the entire system, including thermal decontamination and offgas treatment, would be much higher. In extensive testing at Porton Down, agent vapors were never detected at the entrance to the carbon adsorption bed, let alone the exit (DiBerardo, 2004). Evidently, the offgas cleanup prior to the adsorption beds was effective, and a DRE of at least 99.9999 percent may be assumed.

Unlike the EDS and the DAVINCH, the CDC does not have provisions for holding, testing, and retreating detonation debris before opening the detonation chamber, a feature that many public stakeholders desire.

Public concerns in the United States about using the CDC to treat chemical munitions are not known at this time. However, the extensive U.S. use of the CDC for destruction of conventional munitions, including at the Naval Surface Warfare Center (Bixler, 2005), the Massachusetts Military Reservation, and the Blue Grass Chemical Depot, may contribute to public acceptance. The operations at Blue Grass were conducted under a RCRA permit.[14] The experience with conventional munitions seems to demonstrate that the CDC can be operated without noise or vibration problems for its neighbors.

Secondary Waste Issues

Since the introduction of hot air purging for the CDC system, the secondary waste concerns regarding CDC operations have been substantially reduced. The primary wastes are solids:

- Munition fragments,
- Pea gravel and dust,
- Lime from the reactive bed filter, and
- Carbon from adsorption units.

It was reported that the hot air purging (450°F for 24 hours) yields solids in a condition suitable for transport under government control (Blades et al., 2004). Some post-treatment, such as smelting for metal scrap or incineration for carbon, may be required if the solids are not to be disposed in a hazardous waste landfill.

Process Cost Issues

No quantitative cost information was available to the committee, but some qualitative factors indicate that the CDC technology may be cost effective for some non-stockpile applications. Chief among these factors is the use of the CDC for RCWM destruction operations in Belgium over a period of almost 5 years, including an upgrade in technology from a prototype version of the T-10 model to the more sophisticated TC-60 model.

Similarly, extensive U.S. experience with destruction of conventional and agent-like munitions (smokes, white phosphorus, CS agent) indicates that the basic CDC technology is cost effective for destroying projectiles and other types of explosive-containing munitions in a U.S. context.

Perhaps the most appropriate technology against which to compare cost effectiveness in non-stockpile applications is the EDS-2, which, like the CDC, performs the complete sequence of accessing the chemical agent, destroying the agent, and yielding solid debris that may be disposed of by a TSDF. For small caches of RCWM (one or two munitions), a comparison between the EDS and the T-10 model of the CDC may be appropriate because they appear to be comparable in complexity and mobility. A detailed analysis of costs, including those of waste disposal, would be necessary to see if the CDC offers any advantages over the EDS for sites involving "small finds," i.e., limited numbers of items.

For large caches of RCWM such as may be found at old burial sites, the presumed greater productivity (munitions per week) of the larger CDC systems would seem to offer a cost advantage over the EDS-2. Again, a detailed analysis based on productivity demonstrated in the 2006 Porton Down tests would be required to establish the presumed cost advantage. In this type of operation, the CDC should also be compared to transportable versions of the DAVINCH and Dynasafe systems.

Summary

The CDC system is relatively mature, having been used in a production mode for destroying RCWM in Belgium for more than 4 years in addition to also having been used extensively in the United States for destroying conventional munitions. Modifications made during testing at Porton Down have minimized manual operations and have almost

[13]Brint Bixler, CH2MHILL, responses to committee questions of February 6, 2006.

[14]Meeting between Brint Bixler and John Coffey, CH2MHILL, and committee representatives, Keck Center of the National Academies, Washington, D.C., January 30, 2006.

entirely eliminated the production of liquid wastes. Agent emissions during normal operations appear to have been completely eliminated.

The basic design and operating principles of the CDC are simple. Munitions are encased in explosive and loaded into a large, almost cubical, double-walled steel chamber along with bags of water for thermal control and steam generation. The system is sealed and the explosive is detonated. This explosion breaks open the munition, detonates any energetics contained therein, and releases the chemical agent. The heat, oxygen, and steam in the detonation and expansion chambers destroy over 99.99 percent of the chemical agent. Starting immediately after detonation and proceeding over a 10-15 minute period, the offgases are released to the pollution control system, where they are filtered, the acidity is neutralized, and organic matter is oxidized catalytically. These steps reduce the agent concentration below detection limits before the gases are vented through a bank of carbon adsorption beds. The internals of the destruction systems are decontaminated with hot air, which also decontaminates the residual solids such as munition fragments.

The CDC is safe, reliable, and effective. It is made in three transportable versions that are appropriate for destroying small, medium, and large numbers of munitions. In addition, there is a large fixed model that could be used at a large burial site or firing range.

The smallest mobile CDC model (T-10) seems generally comparable to the EDS-2 in size and complexity. The T-10 has an advantage relative to the EDS in that it produces little or no liquid waste, but it lacks the hold-test-release capability of the EDS for assuring that offgases are devoid of agent emissions. A detailed cost calculation would be required to determine the cost effectiveness of the CDC T-10 vs. the EDS-2 for disposing of small RCWM caches (ones or twos). The presumed greater productivity of the larger CDC models (TC-25 and TC-60) might make them more cost effective for destroying large quantities of RCWM.

The CDC might gain public and regulatory acceptance in the United States without excessive difficulty on the basis of extensive prior operating experience and testing, but some community members may view the lack of a hold-test-release capability as a disadvantage. The committee does not believe that this lack is a significant technical issue, given the batch nature of the process and the proven effectiveness of the offgas treatment system. Still, it believes that this is one of the many factors that must be considered when comparing the CDC with other detonation technologies.

DETONATION OF AMMUNITION IN VACUUM INTEGRATED CHAMBER

Description

DAVINCH is a trademarked acronym for the detonation of ammunition in a vacuum integrated chamber and is a controlled detonation system for the disposal of chemical munitions.[15] DAVINCH technology was developed by the Japanese company Kobe Steel, a manufacturer of large steel pressure vessels. Munitions placed in the DAVINCH vessel are detonated in a near vacuum using a slurry explosive to open the munitions and access the chemical agent. The agent is destroyed as a result of the high temperature (3000K) and pressure (10 gigapascals) generated by the shock wave, followed by high-speed cavitation and then a fireball. DAVINCH is a dry process in that no post-detonation reagent is used because the agent is destroyed in the vessel (see Figure 4-2).

DAVINCH technology is a successor to an explosion containment vessel (DV10) that was used in 2000 at Lake Kussharo on Hokkaido Island in Japan to explosively access 26 World War II bombs containing a mixture of mustard agent and lewisite (Yellow bombs). Holes were drilled in the bombs and the agent was drained and neutralized. The drained bombs, containing explosives, were placed in the DV10 and destroyed using slurry explosives. A successor vessel was developed that was able to both access the agent and destroy it, as noted above. This vessel, the DV45, has been used at Kanda Port in Kyushu Island, Japan, to destroy recovered Yellow bombs and recovered Red bombs containing Clark I and Clark II vomiting agents (DC/DA) (see Figure 4-3). Between October 2004 and May 2005, 100 Yellow bombs weighing 50 kg each and 500 Red bombs weighing 15 kg each were destroyed in the DV45. The experience in using DAVINCH at Kanda Port is described in Lefebvre et al. (2005a), Asahina et al. (2005), and Asahina (2005). A detailed description of the DAVINCH, its design basis, its structural and operational characteristics, and the testing conducted to date are found in Lefebvre et al. (2005b).[16]

The DAVINCH is a double-walled steel chamber. The replaceable inner vessel is made of armor steel and the outer vessel is made of multilayered carbon steel plates with a corrosion- and stress-crack-resistant inner plate made of, for example, stainless steel, Hastalloy, or a similar material. The chambers are separated by air. Owing to its double-wall design and the materials of construction, the DAVINCH has the ability to confine high-pressure detonation gases, eliminating the need for an expansion tank to contain them following a detonation.

The DV45 weighs about 75 tons and has an explosive containment capacity of 45 kg TNT-equivalent. Its inner vessel has an inside diameter of 2.6 meters and an inner length of 3.5 meters. In contrast, the U.S. EDS-2 has a diameter of 0.74 meters and a length of 1.42 meters. A larger version

[15]Except where otherwise noted, the majority of the technical information in this section came from various meetings with representatives of Kobe Steel (Japan) (see Appendix D).

[16]Joseph Asahina, Kobe Steel, "DAVINCH: Detonation of ammunition in vacuum integrated chamber," presentation to representatives of the committee on November 11, 2005.

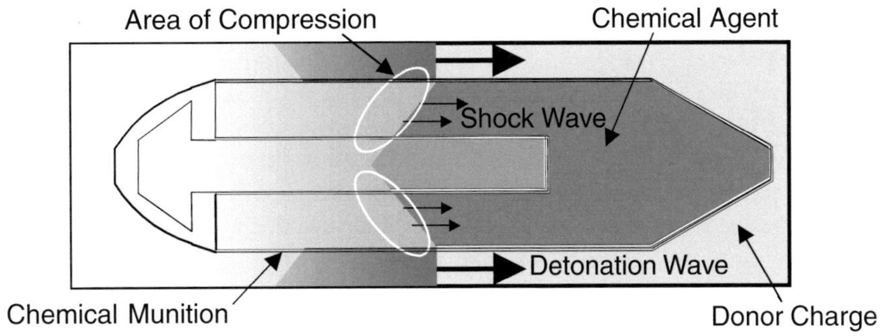

	Chemical Agent Destruction Mechanism
1st step	**Instant compression by propagating shock wave pressure of 10 GPa** (similar phenomenon is observed in cavitation bubbles when bubbles collapse → sonochemistry)
2nd step	**High-speed mixing of chemical agent with detonation gas at high pressure and high temperature**
3rd step	**Thermal decomposition by the long-lasting fireball of 2000°C for 0.5 sec.**

FIGURE 4-2 DAVINCH three-stage destruction mechanism. SOURCE: Joseph Asahina, Kobe Steel, December 8, 2005.

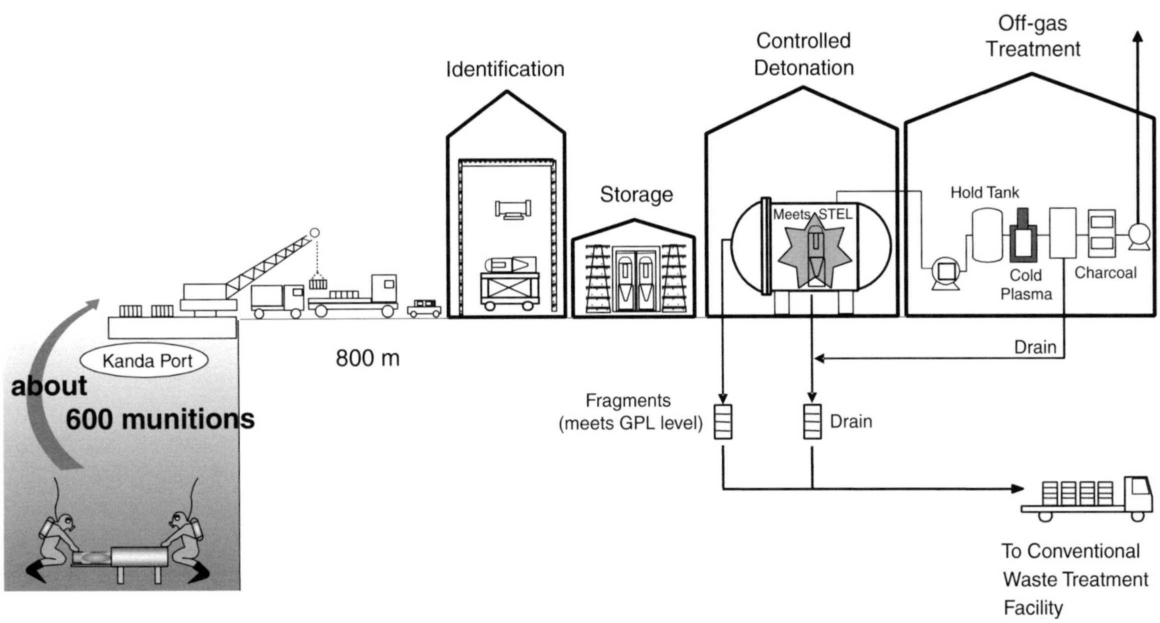

FIGURE 4-3 Outline of the Kanda project. SOURCE: Joseph Asahina, Kobe Steel, December 8, 2005.

of DAVINCH, the DV65, has been fabricated and is available. It has the same diameter as the DV45 but is longer and has an explosion containment capacity of 65 kg TNT-equivalent.

Munitions to be destroyed in a DAVINCH vessel are placed in a box—one munition per box with spacers at each corner to provide room for injecting an emulsive explosive around the munition. The explosive is extruded into the gap between the munition and the inner wall of the box either manually or automatically. The emulsion explosive can also be injected into the overpacks of leaking munitions or, if there is a filler between the overpack and the munition, the explosive can be placed outside the overpack. In this case, additional donor explosive is added to ensure that the explosive in the munition burster is sympathetically detonated by the blast.

A detonator is inserted into the slurry explosive that surrounds the munition and the top of the box and a lifting sling is attached. The munition in its box, with the detonator and detonation wire attached, is lifted by the sling and carried into the vessel by a robotic arm mounted on an operation deck that does not touch the inner walls of the vessel. The robotic arm hangs the sling from a hook on a linear rack at the top of the vessel and then connects the firing wire to a plug-in fixture mounted inside the vessel door. The prongs at the end of the detonation wire are inserted by the robotic arm into a sealed, gas-tight port in the side of the vessel.

The boxed munitions are positioned along the long axis of the vessel a specific distance apart depending on their configuration and contents. The DAVINCH contains an airtight, circular, double-flanged door that is remotely opened and closed. The door is not hinged but moves laterally until it is aligned with the vessel. It is then moved toward the vessel until contact is established and then secured in place. Following a detonation, the door's flanges and gasket can be cleaned using the same robotic arm that moves munitions into the inner vessel.

After the door is sealed, air is evacuated from the inner vessel using a vacuum pump. This process takes about 10 minutes. The resulting vacuum reduces noise, vibration, and blast pressure, thus increasing the vessel life. The munitions are then detonated under near-vacuum conditions (about 0.2 psi). Using an electric delay detonator, the munitions are sequentially detonated such that the second munition is detonated before the shock wave from detonation of the first munition reaches it. The detonations are sequential to reduce the maximum pressure on the inner vessel walls. If more than two munitions are to be sequentially detonated (three have been sequentially detonated in the DV65), the length of the inner vessel can be increased, holding the vessel diameter constant. The munitions are imploded, reducing noise, vibration, fragment velocity, and gouging/scoring of the walls of the inner vessel. By detonating in a near vacuum, the volume of offgas to be treated is also reduced, since following a detonation, the vessel is repressurized to 1 atmosphere and the volume of offgas that is pumped out is the volume of the DAVINCH inner vessel. As a result, an expansion tank is not needed.

The initial shock wave from the detonation of explosives increases the pressure in the inner vessel to up to thousands of atmospheres (10 gigapascals) in 0.3 milliseconds. As illustrated in Figure 4-2, agent is destroyed as a result of a three-sequential-step process:

1. Destruction by a propagating detonation shock wave that compresses the agent.
2. Destruction due to high-temperature and high-pressure detonation gases.
3. Thermal destruction resulting from a 2000°C fireball in the vessel. A proprietary additive increases the time duration of the fireball to 0.5 seconds to ensure agent destruction.

Following the detonation, air is introduced into the inner vessel, with atmospheric pressure reached after about 1 minute. Using the vacuum pump, the internal pressure in the vessel is again reduced to a near vacuum in order to remove the offgases resulting from the detonation of munitions and destruction of agent and energetics. If agent is detected in the offgas, the capability exists to recycle the gas back into the vessel.

Several methods are available to cleanse the DAVINCH vessels. An electrostatically charged decontamination aerosol can be sprayed in the inner vessel and in the gap between the inner and outer vessel in the event that any residual agent is detected. This is done prior to removing the replaceable inner vessel. A water jet spray is available to rinse out this decontamination solution. Finally, following the evacuation of the offgas from the inner vessel, the DAVINCH door can be opened and an explosive cleansing shot can be placed inside. The door is closed and the explosive charge detonated in the empty inner vessel to destroy any residual agent by means of the shock wave and heat from the detonation of the explosive.

Munition fragments are left in the inner vessel and are removed by the robotic arm after a period of time, about once per week. As a result of the heat generated by the fireball, the metal fragments are decontaminated to a point such that they are releasable to the public—that is, they do not exceed the Centers for Disease Control's recommended general population limit (GPL) value for the agents destroyed (for mustard agent, this value is 10^{-6} mg/m^3).

Following the detonation, offgases are cleaned, filtered, and stored in a buffer tank. They are then pumped into a combustion chamber and heated. The combustion gases are quenched and passed through an activated carbon adsorption bed before being released to the atmosphere. An alternative to combustion that is under consideration involves sending the filtered offgas to a small, cold plasma arc unit to treat the gas prior to its release.

TABLE 4-3 DAVINCH Experience in Destroying Japanese WW II-Era Bombs Containing Lewisite, Mustard Agent, and Agents Clark I and Clark II (Vomiting Agents)

Type of Bomb	Length[a] (cm)	Width (cm)	Weight (kg)	Quantity of Explosives (kg)	Quantity of Agent (kg)
Yellow	70	20	50	2.3[b]	18.9
Red	50	10	15	1.3[c]	0.37

[a]Without tail fins.
[b]Picric acid.
[c]TNT-15% naphthalene.

SOURCE: Asahina et al., 2005.

Since the agent in the Yellow bombs destroyed by DAVINCH is a 50:50 mix of lewisite and mustard agent, arsenic removal is part of the process. Eighty percent of the arsenic is metallic and can be recovered without further treatment. The remaining 20 percent consists of arsenic oxides and requires further treatment. The arsenic and arsenic oxides recovered from the DAVINCH process are sent to a contractor for such treatment.

When destroying two Yellow bombs per shot, DAVINCH DV45 has had a throughput rate of three shots per 8-hour day or about 2.5 hours per cycle, including preparation of the munitions, loading the inner vessel, sealing the door, drawing a vacuum, the detonation itself, evacuating the offgases, vessel decontamination if needed, and opening the vessel door to prepare for the next cycle. The DV65 has processed up to three Yellow bombs per shot, or nine bombs per 8-hour day. Under automatic operation, the throughput is estimated to be five shots per 8-hour day according to the manufacturer, but this has yet to be demonstrated.

Country-by-Country Experience

The DAVINCH DV45 unit has been used in Japan, at Kanda Port, to dispose of 600 World War II chemical bombs, some containing a lewisite/mustard agent mix and others containing vomiting agents (Lefebvre et al., 2005a; Asahina et al., 2005; Asahina, 2005). This is the only use of DAVINCH technology to date. It is expected that this DAVINCH unit will be used again at Kanda Port to dispose of bombs that remain underwater and that will be brought to the surface in sealed containers. It is also possible that DAVINCH technology will be used for applications in China, France, and Belgium in the future.

Evaluation Factors Analysis

Process Maturity

DAVINCH is a developed technology with experience in destroying Japanese World War II-era bombs containing lewisite, mustard agent, and agents Clark I and Clark II (vomiting agents). The characteristics of these bombs are given in Table 4-3.

Although DAVINCH technology has not been tested or used with U.S. non-stockpile munitions, the stated capabilities indicate it could process such munitions. For example, a representative large non-stockpile item, the 8-inch, T-174 projectile, has a length of 35.17 inches (89 cm), a width of 8 inches (20.3 cm), and weighs 200 pounds (91 kg)—somewhat longer and heavier than the Japanese Yellow bomb but still within the physical capability of a DAVINCH DV45. This projectile contains 6.95 pounds (3.15 kg) of Composition B explosive in its burster, a somewhat greater quantity than found in the Japanese Yellow bomb but still well within the 45 kg explosive containment capability of the DV45.

Although DAVINCH technology is used in Japan, it has not been permitted for use in the United States, but the manufacturer, through a U.S. corporate partner, is looking into permitting requirements and procedures. As of the close of information gathering for this report, Kobe Steel has not yet applied for a permit to test DAVINCH technology in the United States.

Process Efficacy/Throughput

DAVINCH technology appears to be well suited for destroying a variety of non-stockpile munitions and containers in the United States as well as for destroying both stockpile chemical munitions and conventional high-explosive rounds (although it has not been used to destroy munitions filled with nerve agents). It has the potential to destroy chemical weapons with different fills in a single shot as well as to destroy a combination of chemical and conventional munitions in a single shot, although this has yet to be demonstrated.

DAVINCH units exist that are considerably larger than the largest detonation vessel used by the U.S. Army's NSCMP, the EDS-2. The DAVINCH vessel used in Japan at Kanda Port, the DV45, has an inner diameter of 2.6 meters and an inner length of 3.5 meters. Comparable dimensions for the EDS-2 are 0.37 meters and 1.42 meters; consequently, the internal volumes differ by a factor of about 30. Explosion containment capabilities are also substantially different: 45 kg (99 pounds) for the DV45 vs. 5 pounds for the EDS-2, a factor of about 20. The DAVINCH footprint, including the detonation vessel, gas treatment, lab space, and personnel support, is a rectangle having dimensions of about 80 meters by 60 meters, based on the Kanda Port experience.

DAVINCH units can be mounted on a flatbed trailer and made transportable; this is planned for use at various locations in China, where relatively small quantities of munitions have been found. At these locations, a transportable unit is more cost effective than construction of a fixed facility. Supporting infrastructure would also be transportable.

Kobe Steel has estimated the DRE for the detonation chamber at >99.9999 percent. However, the procedures were not consistent with U.S. regulatory requirements—that is, the methodology cannot be used to calculate the regulatory DRE. The committee believes, however, that the DAVINCH technology should be able to achieve a high DRE, considering that no agent has been detected downstream of the detonation chamber.

The gases resulting from detonation in the DAVINCH vessel are primarily H_2 and CO. These gases are pumped from the vessel and passed through a cyclone to remove particulates. They are then held in a storage tank for testing of the offgas content. In the event that 99.9999 percent DRE is not achieved, the offgas can be returned to the DAVINCH vessel for further treatment via a cleansing shot in which another detonation takes place. Rather than being returned to the vessel, the offgas can also be sent to an adjacent combustor and passed through a two-bed charcoal filter before being released to the atmosphere.

As an alternative to offgas combustion, the DAVINCH manufacturer is considering use of a small cold plasma unit to treat the detonation offgas. The cold plasma unit is about 1.5 meters high and has the appearance of a home hot water heater—basically a vertical cylinder. The unit operates at a temperature of 900°C and processes about 1 m^3 of offgas per minute, based on the 20 m^3 offgas volume resulting from a shot in the DV45 and a 20-minute processing time to pass the gas through the cold plasma unit. It operates under a slightly negative pressure, and an oxygen supply is provided to aid in the destruction of the offgas constituents. Although intended to be a gas treatment unit, the cold plasma also is claimed to remove 99.9 percent of any agent that may remain in the offgas. This unit can be plugged into a standard 220-volt wall outlet.

With the cold plasma unit as an alternative for offgas processing, a proposed modification to the process flow would place the plasma unit before the offgas storage tank. The treated offgas can still be held in the tank and tested for its constituents. If any agent is detected, the treated offgas can be returned to the DAVINCH vessel for further treatment via the cleansing shot or can be recirculated through the vessel and returned to the cold plasma unit for further treatment in that unit.

At Kanda Port, the DAVINCH DV45 processed two Yellow bombs per shot with an average cycle time of 150 minutes, or 3.2 shots per 8-hour day. Over a 3.5-month period, 600 bombs were destroyed in 250 shots; an average of 2.4 bombs per shot. Assuming a 22-working-day month, the average number of shots per day was $250/(3.5 \times 22) = 3.25$, consistent with the 150-minute cycle time per shot.

Each Yellow bomb contained 18.9 kg of lewisite/mustard agent fill; thus, 83.3 pounds of agent were destroyed per shot. If manual operations, e.g., inserting the emulsion explosive into the box containing the munition, are replaced with a more automated operation, the DAVINCH throughput may increase to five shots per day, although this has not been demonstrated. Also, a larger version of DAVINCH (DV65) has the capability of destroying three Yellow bombs in a single shot; thus with automated operation, a throughput of up to 15 munitions per day is possible.

For U.S. non-stockpile munitions, the expected throughput will depend on several factors, including the size of the DAVINCH vessel to be used, the munition size, the quantity of agent to be destroyed, the explosive content of the munition and the donor charge, and whether or not automated handling procedures are used. Estimated throughput rates per 10-hour day for representative U.S. munitions have been provided by Kobe Steel and are shown in Table 4-4. These rates are for a DAVINCH DV65 having an explosive containment capability of 65 kg TNT-equivalent and assume that manual handling procedures are used. If automated procedures were to be used, the estimated number of cycles per 10-hour day would increase from 6 to 8.

The cycle time that was provided by the technology proponent for the DV65 operating under manual handling procedures was 1.5 hours. This is equivalent to the 6 cycles per day given in Table 4-4 plus a presumed 1-hour allowance for start-up and shutdown and/or minor delays. This cycle time is substantially shorter than the demonstrated 2.5-hour cycle time for the smaller DV45 that operated in Japan.

The quantity of agent that can be destroyed in a single DAVINCH cycle will also vary. Table 4-5 gives these quantities for the same munitions as those in Table 4-4.

Because there is no neutralization step, the throughput rate for DAVINCH is higher than it is for the EDS-2, which conducts only one detonation every other day, albeit with up to six munitions destroyed per detonation, depending on the munition size. The DAVINCH generates some liquid wastes. These result from use of the decontamination spray, when used; from residual liquid in munitions recovered from underwater; and from the cooling of the offgas. The volumes are small relative to those generated from neutralization and are sent to an offsite waste treatment facility for further processing and disposal.

TABLE 4-4 Estimated DAVINCH DV65 Throughput Rates

Munition	Munitions per Cycle	Cycles per 10-hr Day	Munitions per 10-hr Day
4.2-in. mortar, M1	6	6	36
75-mm projectile, M64	5	6	30
5-in. projectile, MK VI	3	6	18
5-in. projectile, MK 54	2	6	12
155-mm projectile, MK II	2	6	12
8-in. projectile, T174	1	6	6
Bomblet, M139	12	6	72
105-mm projectile, M60	5	6	30
100-lb bomb, M47	1	6	6
115-lb bomb, M70	1	6	6

SOURCE: Information provided by Ryusuke Kitamura, Kobe Steel, Ltd., to the committee, March 25, 2006.

TABLE 4-5 Agent Quantities Destroyed per DAVINCH DV65 Cycle

Munition	Agent and Weight	Items per Cycle	Agent Weight per Cycle (lb)
4.2-in. mortar, M1	Mustard agent, 6.5 lb	6	39
75-mm projectile, M64	Mustard agent, 1 lb	5	5
5-in. projectile, MK VI	Mustard agent, 5.4 lb	3	16.2
5-in. projectile, MK 54	GB, 4.2 lb	2	8.4
155-mm projectile, MK II	Phosgene, 11 lb	2	22
8-in. projectile, T174	VX, 15.7 lb	1	15.7
Bomblet, M139	GB, 1.3 lb	12	15.6
105-mm projectile, M60	Mustard agent, 3.2 lb	5	16
100-lb bomb, M47	Mustard agent, 70 lb	1	70
115-lb bomb, M70	Lewisite, 83 lb	1	83

SOURCE: Information provided by Ryusuke Kitamura, Kobe Steel, Ltd., to the committee, March 25, 2006.

The inner DAVINCH vessel is replaceable and, as stated by the manufacturer, can be used for at least 1,000 shots. Because the munition fragments tend to strike the vessel walls in the same general area following each shot, the liner is periodically rotated in order to distribute the impact areas around the circumference of the vessel.

Process Safety

DAVINCH requires between 20 and 25 workers plus laboratory personnel. All operations involving munition handling and the manual insertion of slurry explosive around the munitions are carried out by workers wearing low-level PPE (Level D). Higher levels of PPE are used if leaking munitions are to be handled. Since insertion of the munitions into the inner vessel is done using a robotic arm, presumably there is no worker exposure during that operation. Following detonation and evacuation of offgases, a spray decontamination solution is used if residual quantities of agent are detected. The heat-treated munition fragments are periodically removed remotely. Consequently, there should be no worker exposure to agent after the munitions are destroyed.

Public and Regulatory Acceptability in a U.S. Context

DAVINCH technology has not been permitted for use in destroying chemical weapons in the United States, although it has been used successfully in Japan for this purpose. No significant regulatory issues were identified to indicate that the DAVINCH technology could not meet U.S. environmental regulatory requirements if appropriate information (such as verified DRE, residual levels of dioxin, furans, arsenic, and any other chemicals of regulatory concern) is developed and provided to the regulators in a timely manner.

Additional testing of DAVINCH technology will be required prior to its being permitted in the United States for treatment of chemical weapons and materiel. Following a detonation, the inner vessel can be monitored for the presence of agent and, if necessary, an additional explosive cleansing shot can be carried out to remove trace quantities of agent, and/or a spray decontamination solution can be injected into the inner vessel for the same purpose. Offgases from the detonation are held in a storage tank and tested for agent. Depending on the agent level detected in the offgas, it can be either returned to the inner vessel for further agent destruction in a cleansing shot or sent to a gas treatment unit—either a combustor (incinerator) or a cold plasma unit—before being passed through carbon adsorption beds.

The public reaction to DAVINCH is not yet known and may be complex since no DAVINCH units are operating in the United States upon which to base a perception. Moreover, there is no U.S. regulatory experience with this technology, and the use of thermal treatment to destroy any remaining agent in the detonation offgases might receive a mixed reaction from both the concerned public and regulators. However, public acceptance is likely to be favorable in light of the high DRE that is achieved and because all process residuals can be held and tested prior to release. Moreover, because the munitions are detonated in a vacuum, DAVINCH technology can be used in an urban area (and was so used in Japan) with greatly reduced noise and vibration, possibly to a point where these would not be of concern to the general public. At a distance of 0.2 km (640 feet), the noise resulting from a DAVINCH detonation was reduced from 72 dB at atmospheric pressure to 65 dB under vacuum conditions, and this 7 dB reduction in noise held for greater distances as well. An extensive public outreach process was undertaken prior to and during use of DAVINCH at Kanda Port in Japan, with frequent meetings held with public interest groups (Asahina, 2004). It is anticipated that a similar outreach effort would take place in the United States were DAVINCH technology to be used here.

Secondary Waste Issues

The waste streams produced by the DAVINCH technology are (1) gases resulting from the detonation and (2) heat-treated munition fragments that have been decontaminated to a point where they can be released or recycled. The gases can be stored in a buffer, tested for agent and other constituents, and sent to a post-processing facility for cleaning. Although the gases are currently combusted/incinerated and scrubbed, it may be possible to treat them in a plasma arc process that would clean them and destroy any residual agent. The public acceptability of doing this is not known, since treatment in a plasma unit could also be perceived as incinerating the offgases.

Arsenic recovery also presents a problem since nearly all of the arsenic resulting from DAVINCH operations is in dust, on munition fragments, or on the walls of the inner vessel. Although most of the arsenic on the vessel walls can be scraped off, some may remain in microcracks in the vessel wall that result from the detonations. Because removal of this arsenic is difficult, it is not routinely removed.

Process Cost Issues

Quantitative cost information for the acquisition and operation of a DAVINCH system was not available to the committee. Based on operating experience in Japan, the DAVINCH could be a cost-effective technology, especially if moderately large quantities of items (several hundred or more) are to be destroyed and if the physical sizes and/or the net explosive weights of the items to be destroyed exceed the capacities of other detonation-based technologies.

Operating costs may be greater than they are for the EDS since more staff may be needed (about 20 to 25 for the DAVINCH vs. 6 to 12 for the EDS). This may be offset, however, by the fact that DAVINCH technology has a greater capacity for accepting munitions and a higher throughput rate than the EDS, thus shortening the time that may be required for a specific application.

The life-cycle costs of acquiring, installing, operating, and removing a DAVINCH unit at a particular location will depend on numerous factors, including (1) the costs of acquiring the DAVINCH unit and transporting it and related equipment to the site; (2) site preparation costs; (3) the number of items to be destroyed, their explosive configuration, and the quantities of agent fill (these factors will influence the throughput rate and time duration of a campaign); (4) site-specific regulatory compliance costs; (5) the costs of secondary waste treatment; and (6) the requirements for disposal of treated residuals.

Summary

The DAVINCH technology uses a large detonation chamber in which chemical munitions and their contents are destroyed when donor charges surrounding the munitions are detonated under a near vacuum. Although the process does not require use of a reagent to destroy the agent—the destruction is accomplished by a shock wave, expansion and thermal heating from the detonation gases, and a fireball in the chamber—offgases are produced that require some secondary treatment by, for example, combustion and scrubbing.

DAVINCH technology has been used in Japan to destroy 600 Japanese chemical bombs, some containing a lewisite/mustard agent mixture and others containing vomiting agents. The technology has not been used to destroy any U.S. non-stockpile chemical munitions.

The size and the explosion containment capability of versions of the DAVINCH technology are substantially greater than those of the largest treatment technology used in the United States for RCWM (the EDS-2), and its throughput also exceeds that of the EDS-2 by a factor of at least 3. It has demonstrated the ability to destroy over 80 pounds of agent (a lewisite/mustard agent mix in two Japanese Yellow bombs) in a single application and to have destroyed 10.14 pounds of explosive (picric acid) in these bombs.

The DAVINCH technology appears to be safe and effective. The external donor charges allow DAVINCH to be used to open agent-filled containers, inert munitions, and munitions containing energetics in order to access and destroy the agent. Because it is larger, DAVINCH is less mobile than the EDS-2, although a transportable version is under development.

TABLE 4-6 Size Specifications for Two Dynasafe Static Kiln Models

	SK1200	SK2000
Explosive containment TNT-equivalent, lb (kg)	2.64 (1.2)	5.06 (2.3)
Length, m	4.5	6.0
Width, m	4.35	5.5
Height, m	6.0	8.0
Weight, kg	24,000	40,000
Approx. detonation chamber volume, m^3	0.91	4.19

SOURCE: Information provided to the committee by UXB International, Inc., August 19, 2005; <http://www.dynasafe.com/destruction-of-munitions-static-kiln.html>.

Although application of DAVINCH technology to future U.S. non-stockpile disposal needs will depend on the nature of the items to be disposed of, DAVINCH technology has potential applicability at those U.S. sites where a temporary facility can be placed and could be used to dispose of medium to large quantities (hundreds to thousands) of items containing chemical agent or that are agent contaminated. It is probably not cost effective to dispose of items unlikely to contain agent, e.g., containers that have been previously burnt out, or for small numbers of small chemical-containing items, e.g., bomblets or small caliber projectiles, where the EDS technology would have greater applicability.

DYNASAFE TECHNOLOGY

Description

Dynasafe is the tradename for a static kiln manufactured by Dynasafe AB, a Swedish company that designs and manufactures products for the containment of explosions, including mobile explosion containment vessels used by police departments and the Burster Detonation Vessel, used by the NSCMP at its Munitions Assessment and Processing System facility in Edgewood, Maryland.[17]

The Dynasafe static kiln is a near-spherical, armored, dual-walled high-alloy stainless steel detonation chamber (heated retort) inside a containment structure (Ohlson et al., 2004).[18] The total thickness, including a safety layer, is 15 cm. The detonation chamber can operate in a pyrolytic or oxidizing environment. Intact munitions are indirectly heated by electrical resistance elements between the inner and outer walls of the detonation chamber. The munitions are heated to a temperature of 400°C-600°C, resulting in deflagration, detonation, or burning of the munition's explosive fill. The chemical agent in the munition is destroyed as a result of the shock wave from the detonation when this occurs, the resulting gas pressure (measured at 10 bars, or 9.87 atmospheres), and decomposition due to the heat in the chamber. No explosive donor charge is used, and no reagent is needed to neutralize the agent. The kiln operates in a semibatch mode. Two sizes of the static kiln are available. Specifications are provided in Table 4-6.

Chemical munitions are placed in a cardboard box or carrier, preferably by robot but if need be, manually. The box is placed on an elevator for the SK2000 version or on a trolley conveyor for the smaller units and is transported to the top of the kiln. Leaking munitions are placed in an airtight plastic bag and then in the box before being loaded. Munitions that are already in a single round container can be loaded onto the conveyor or elevator while in the container.

The boxed munitions are fed into the kiln through two loading chambers (see Figure 4-4), each having its own hydraulically operated door and inflatable seal. The upper loading chamber has airlock doors and the lower loading chamber has a hot blast door between it and the kiln's detonation chamber. The doors, loading chambers, and detonation chamber are all designed to resist and contain the overpressure from a detonation of up to 2.3 kg TNT-equivalent. An additional 2.3 kg TNT-equivalent of overpressure containment is included in the design as a safety margin. To provide total containment, the doors are gas-tight as well as explosion-resistant. The interior of the detonation chamber is not open to the atmosphere while munitions are loaded, and the loading chambers are offset for safety purposes.

Using a hydraulic arm, the boxed munitions are pushed into the loading chambers, moving from one chamber to another, and are then dropped onto a heated (500°C-550°C) shrapnel (scrap) bed at the bottom of the detonation chamber. The maximum drop is about 2 meters. The purpose of this bed is to protect the chamber walls from munition fragments when detonation occurs. If sufficient energy from energetics in the munition is released, no additional external heating from the electrical resistance elements is required. If the munition does not contain energetics, then additional heat can be provided by the electrical resistance elements.

[17]Except where otherwise note, technical information for this section came mostly from meetings with representatives of Dynasafe AB (Sweden) and UXB International, Inc. (United States) (see Appendix D).

[18]See also <http://www.dynasafe.com/destruction-of-munitions-static-kiln.html>.

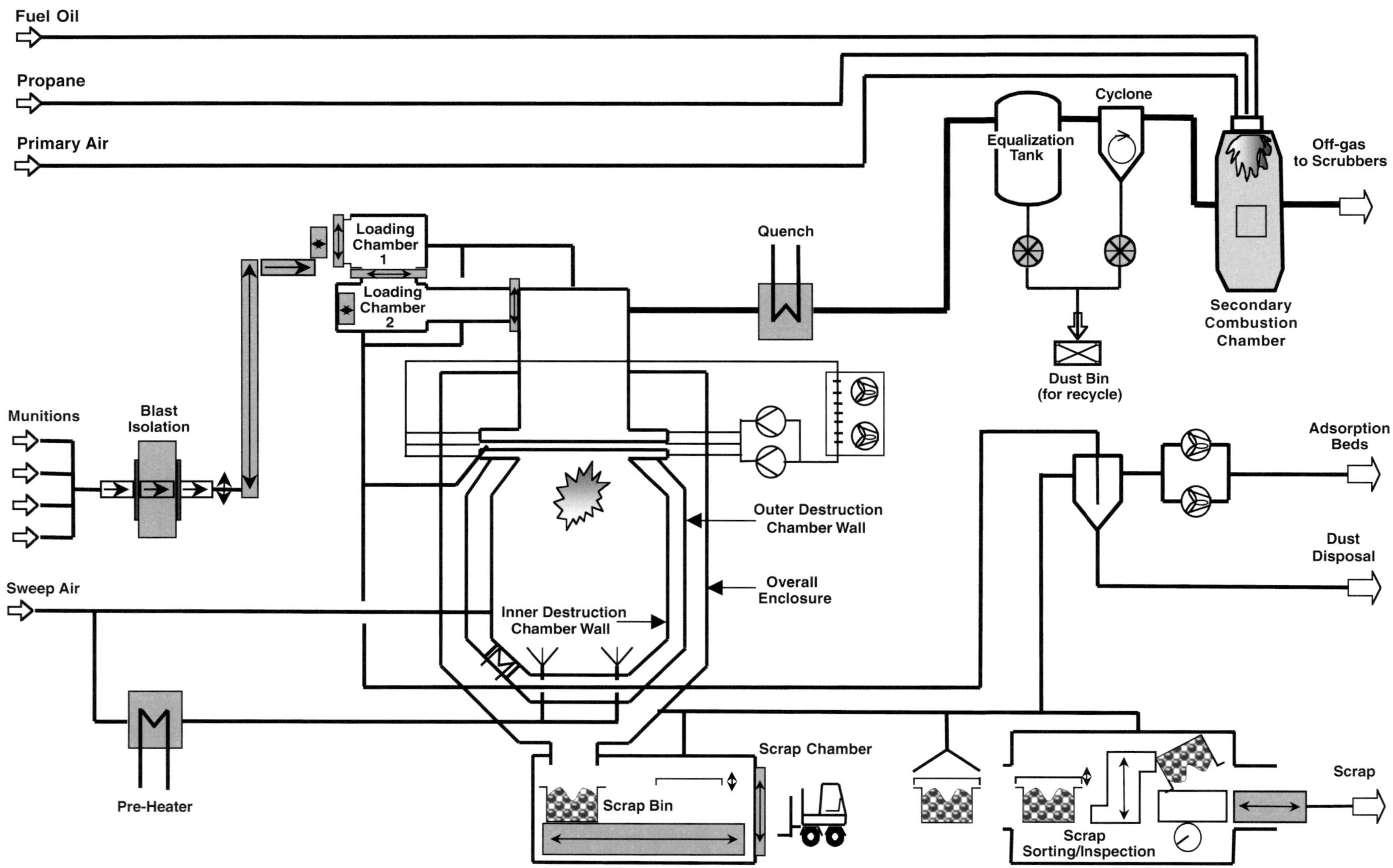

FIGURE 4-4 Dynasafe static destruction kiln process flow. SOURCE: Harley Heaton, UXB International, Inc., April 10, 2006.

During operations, conditions in the detonation chamber are monitored using an air-cooled camera located in a tube that protrudes into the chamber. A slight negative pressure is maintained in the chamber to enable detection of the pressure pulse that takes place when a munition detonates. A microphone is used to detect the sound of a detonation, and vibration of the chamber is also recorded.

When the detonation chamber has a full scrap load, i.e., when it is about 50 percent full, a clean burning period takes place during which the scrap metal is heated to 550°C-650°C for several hours to meet GPL requirements. After completion of the clean burning period, the detonation chamber disengages from the lower loading chamber and is rotated almost 180 degrees clockwise in order that most of the munition fragments can be dropped into a scrap bed in a bin. A low baffle plate in the detonation chamber, near the place where the scrap exits, retains some of the scrap/shrapnel for the next load. The metal scrap bins are enclosed within the outer housing of the kiln to prevent dust from escaping and to allow confirmation that the metal can be released. When scrap removal has been completed, the kiln rotates back to its upright position and the retained scrap in the detonation chamber falls to the bottom.

During operations, offgases from pyrolysis and detonation are continuously evacuated from the kiln, and compressed air is used to sweep all offgases from the combustion chamber. If the process is operated as a closed system—that is, as a batch reactor—the offgases can be held inside the detonation chamber for as long as necessary to ensure that agent destruction takes place. The offgases can also be analyzed prior to their release to the offgas treatment system. If necessary, nitrogen can be used as the sweep gas. When the process is operated as an open system, the offgases are transferred to a heated buffer that serves as an expansion tank and as a cyclone to remove coarse dust. European Union environmental regulations require that to ensure agent destruction, a secondary combustion chamber with a 2-second residence time and operating at 1100°C must be used. Other offgas treatment steps may include use of a quench tower to cool the gases to prevent dioxins and furans from forming, as well as various scrubbers and equipment to capture fine particulates and to remove heavy metals and metallic oxides. The use of such equipment will depend on whether the Dynasafe unit is operated as an open or a closed system, the constituents of the offgas, and environmental requirements.

The elapsed time for a munition destruction cycle will vary with the explosive and agent content of the munition. For conventional munitions, throughput of 25-35 detonation cycles per hour has been demonstrated for explosive loads of 2 kg TNT-equivalent and can be greater for smaller explosive loads. Daily throughput includes the clean burning time. The throughput for chemical munitions will depend on whether the Dynasafe is operated as an open or a closed system, the number of munitions that are fed into the detonation chamber per cycle, and the number of cycles per hour.

Country-by-Country Experience

Dynasafe static kilns have been used to destroy a substantial variety of conventional munitions in several countries. The applications include these:

- Sweden, destruction of detonators and small arms ammunition in SK400 (1997) (no longer available).
- Spain, destruction of conventional munitions in SK1200 (1997).
- Sweden, destruction of conventional munitions in SK800 (1999) (no longer available).
- Japan, destruction of antipersonnel mines and conventional munitions in SK1200 (2000).
- Portugal, destruction of antipersonnel mines and conventional munitions in SK1200 (2001).
- Asia, destruction of conventional munitions in SK2000 (2003-2004).

A prototype development unit has destroyed over 100 kg of mustard, lewisite, and Clark I and II agents, although these agents were not contained in chemical munitions. In February 2006, 100-mm German grenades containing energetics and 1.5 kg of mustard agent fill were successfully destroyed in the Dynasafe SK2000 at the GEKA facility in Munster, Germany.[19] Three grenades were destroyed per feed cycle. The ability of Dynasafe to access and destroy agent in thick-walled steel munitions will also be demonstrated at GEKA. A detailed description of the use of the Dynasafe SK2000 at the facility is provided in Weigel et al. (2004).

Evaluation Factors Analysis

Process Maturity

The Dynasafe family of static kilns is a mature technology that has been used for several years to destroy a substantial variety of conventional munitions, as noted above. The kilns have been both safe and effective for this application. Using this experience as a basis, the Dynasafe static kiln has been modified to destroy chemical munitions and was doing so at the above-mentioned German government facility in Munster, Germany, when this report was being prepared. As of April 21, 2006, at least 1,000 munitions containing mustard agent, phosgene, or diphenylchloroarsine (Clark I) agent had been destroyed.

Modifications include making the kiln gas-tight to contain any agent remaining in offgases, heating the scrap metal to remove all traces of agent on metal surfaces, and using an elaborate offgas treatment system to scrub the detonation gases and remove any remaining traces of agent.

[19]GEKA, Gesellschaft zur Entsorgung von chemischen Kampfstoffe und Rüstungs-Altlasten.

Although the Dynasafe static kiln has not yet been tested or used to process U.S. non-stockpile chemical munitions, it appears to have the capability to do so since many of these munitions are within the size and explosive containment capabilities of the largest Dynasafe unit, the SK2000, and contain the same mustard agent fill found in the munitions being destroyed in Munster. As this report was being prepared, none of the Dynasafe kilns had been permitted for operation in the United States for the destruction of chemical munitions.

Process Efficacy/Throughput

The Dynasafe static kiln heats munitions until the energetics within them detonate, causing the agent to be exposed to the resulting shock wave, blast pressure, and heat. It is possible, however, that for some items, the energetics and/or agent will undergo deflagration (rapid combustion driven by heat transfer). In fact, deflagration rather than detonation is stated to be the usual destruction process in the detonation chamber.[20] Some items only contain agent, the energetics having been removed or never having been placed in the munition (as would be true, for example, with a test round). In these cases, although the agent may vaporize within the munition body and may rupture the munition body as a result, this is not guaranteed to happen. In such cases, the manufacturer states that the agent will escape as it vaporizes, either through the threads in the munition nose closure or through a weak point in the munition body.

In testing at GEKA in early 2006,[21] empty inert grenades were filled with water, welded shut, and placed in the SK2000 detonation chamber. The water fill vaporized and, as a result of the increased internal pressure, destroyed the grenades, as observed by the control room operators. In additional testing, partially sealed, water-filled grenades were placed in the detonation chamber and heated. As internal pressure slowly increased, the water vapor escaped through screw threads. Absent the sudden destruction of the grenades, it was not possible to detect the escaping vapor, and the grenades emerged intact. The grenades were then x-rayed and cut open to verify that they were empty.

Results to date indicate that the agent in all sealed or partially sealed inert munitions is destroyed, although operating results for grenades and other munitions that may contain mustard agent heels were not available. However, the absence of a positive indication that agent destruction has taken place for those munitions where agent slowly escapes may be a concern, and it may increase process costs and complexity if post-processing actions are required to confirm that no agent remains in the munition.

[20]Meeting between representatives of DYNASAFE AB and a committee fact-finding group, Munster, Germany, January 16, 2006.

[21]Holger Weigel, Dynasafe Germany, presentation to the committee on March 1, 2006.

Finally, testing of explosively configured munitions containing agent simulants has been conducted to demonstrate accessing and destruction of the agent simulant in the munitions.

The technical director at GEKA has stated that the worst case would be one in which a munition containing neither agent nor energetics is fed into the chamber: in that case, the munition would experience nothing other than being heated and would emerge as it entered and have to be opened under controlled conditions to ascertain its original condition.[22] Opening the munition would increase costs as well as the potential for human exposure. If processing needed to stop while the munition was examined to confirm that it is empty and inert, throughput might also be reduced.

The Dynasafe static kilns and related material handling equipment are large: For example, the largest unit, the SK2000, is 6 meters long, 5.5 meters deep, and 8 meters high. The weight of this unit is 44.1 tons. A smaller version, the SK1200, is 4.5 meters long, 4.35 meters deep, and 6 meters high. This unit weighs 26.4 tons, but a mobile version is under development (Dynasafe, 2006). The mobile version consists of eight containers: three for the static kiln, three for the offgas treatment system, and two for spare materials and a workshop. These containers can be carried on three flatbed trailers, and the mobile version can be operated in either an open or closed mode.

The explosion containment capabilities of the Dynasafe static kilns are comparable to those of the EDS-1 and EDS-2 in use by the U.S. Army: 2.64 pounds TNT-equivalent for the SK1200 vs. 3 pounds for the EDS-1 and 5.06 pounds TNT-equivalent for the larger SK2000 vs. 5 pounds for the EDS-2. The detonation chamber of the SK2000 is substantially larger than the EDS-2 chamber; it has the approximate shape of a 2-meter-diameter sphere and, thus, a volume of about 4.2 m^3 compared to a volume of 0.61 m^3 for the EDS-2. The largest munition that can be fed into the feed system of the SK2000 currently in operation at Munster is 30 cm in diameter and 60 cm long. The manufacturer states that the feed system can be reconfigured to allow larger munitions, e.g., 8-inch projectiles having a length of 89.4 cm, to be fed through the loading chambers and into the detonation chamber if the need arises.

In the event that larger items are recovered by the NSCMP (such as 100-pound, 500-pound, and 750-pound bombs), their treatment is more problematical because they are all more than a meter long and contain significant quantities of agent. For example, a 100-pound M47 bomb contains 70 pounds of mustard agent and a 750-pound MC-1 bomb contains 220 pounds of sarin (GB). Although these items can be processed through the SK2000, the technology provider states that the amount of agent in these items would require

[22]Hans-Joachim Grimsel, technical director, GEKA, in a meeting with a fact-finding group of the committee, Munster, Germany, January 17, 2006.

TABLE 4-7 Estimated Dynasafe SK2000 Throughput Rates[a]

Munition	Munitions per Cycle	Cycles per Hour	Munitions per Hour	Munitions per 10-hr Day
4.2-in. mortar, M1	4	3	12	120
75-mm projectile, M64	9	3	27	270
5-in. projectile, MK VI	4	3	12	120
5-in. projectile, MK 54	3	3	9	90
155-mm projectile, MK II	2	2	4	40
8-in. projectile, T174[b]	1	2	2	20
Bomblet, M139	16	3	48	480
105-mm projectile, M60	4	3	12	120

[a]Based on operation as an open (continuous mode) system versus a closed (batch mode) system.
[b]A fragment shield would be placed around the body of the 8-inch projectile to protect the detonation chamber walls.

SOURCE: Harley Heaton, UXB International, presentation to the committee on February 15, 2006.

TABLE 4-8 Agent Quantities Destroyed per Dynasafe SK2000 Cycle

Munition	Agent and Weight	Items per Cycle	Agent Weight per Cycle (lb)
4.2-in. mortar, M1	Mustard agent, 6.5 lb	4	26
75-mm projectile, M64	Mustard agent, 1 lb	9	9
5-in. projectile, MK VI	Mustard agent, 5.4 lb	4	21.6
5-in. projectile, MK54	GB, 4.2 lb	3	12.6
155-mm projectile, MK II	Phosgene (CG), 11 lb	2	22
8-in. projectile, T-174	VX, 15.7 lb	1	15.7
Bomblet, M139	GB, 1.3 lb	16	20.8
105-mm projectile, M60	Mustard agent, 3.2 lb	4	12.8

SOURCE: Harley Heaton, UXB International, presentation to the committee on February 15, 2006.

that the bulk of the agent be removed from the ordnance before treatment. The drained agent and ordnance item would be treated separately. The method to be used for agent destruction is not specified.[23]

The demonstrated throughput for the SK2000 processing conventional munitions has varied with the explosive loading. For a load of 4.4 pounds (2 kg) TNT-equivalent, the SK2000 can accept at least 20 loads per hour, a cycle time of 3 minutes per load. The throughput rate for operation with chemical munitions will be less and will depend on how the Dynasafe is operated, the explosive loading, and the composition and quantity of agent to be destroyed. If operated as a closed system with the offgas held and tested prior to release to the offgas treatment equipment, then one cycle per hour is expected. If operated as an open system, then two to three cycles per hour are expected.

The number of munitions fed per cycle will depend on the munition size, the quantity of agent to be destroyed, and the explosive content (net explosive weight). Estimated hourly throughput rates for some munitions have been provided by Dynasafe representatives and are shown in Table 4-7. These rates are for a Dynasafe SK2000 operating in a continuous mode.

The quantity of agent that can be destroyed in a single cycle will also vary. Table 4-8 gives these quantities for the same munitions listed in Table 4-7.

The average throughput rate will include the periodic multihour clean-burning period, when munitions are not fed into the detonation chamber, and the scrap metal in the bottom of the chamber is heated to 550°C-650°C to meet general population limit (GPL) requirements. Dynasafe is capable of handling mixed loads as long as the explosive containment capacity of the detonation chamber is not exceeded. The DRE for chemical agent destroyed in Dynasafe kilns and postprocessing units has been measured at 99.9999 percent and greater, down to the limit of detection for the instruments used. This DRE was demonstrated in a subscale model of the detonation chamber at the GEKA facility in Munster in 2002. Up to 5.5 pounds per hour of mustard agent was destroyed, as well as Clark I and Clark II vomiting agents and $AsCl_3$, with 220 pounds of these agents destroyed under pyrolytic conditions. This prototype, however, was not a blast chamber, and apparently the agents were destroyed by

[23]Information provided by UXB International in response to committee questions of February 2006.

heating and gasifying them in the chamber. The fate of the arsenic in the agent was not specified.

Although agent destruction was demonstrated, the agent was not contained in real or simulated munitions and energetics were not present. Tests of the Dynasafe detonation chamber using nerve agents have also not been conducted and are not planned since these agents are not present in the German chemical items to be destroyed at GEKA.

As noted above, pyrolysis in the detonation chamber is to be followed by offgas treatment, including, as needed, a cyclone, a combustion chamber, a quench tank, and various scrubbers and filters. This offgas treatment process, although standard, is fairly complex when compared to other detonation-based technologies, and its reliability, cost, and effectiveness when processing chemical munitions needs to be demonstrated. It should be noted that this extensive offgas treatment is specific to the Dynasafe installation in Munster, Germany, where a substantial variety of agent fills are anticipated and where the operator wishes to be able to process every expected gas constituent. For a Dynasafe operating in the United States where agent fills may differ and where the regulatory requirements for secondary waste processing may not be the same as the requirements in the European Union, the offgas treatment facility configuration may differ and could be either more or less elaborate than at the facility in Munster depending on the agent fill and on whether the Dynasafe operates as a closed (batch) or an open (continuous) system.

The Dynasafe static kiln and its related equipment take about 3 months to assemble once the equipment is on site. Following its use, the installation takes about three months to disassemble. While in operation, four to eight people are needed to operate the unit: control room staff, a loading supervisor, and an on-call engineer. For operations with chemical items, more staff may be needed, but the number was not available to the committee.

Process Safety

The potential for worker exposure to agent is about the same as with any other operation where RCWM need to be handled, boxed or packaged, and moved. Dynasafe workers do not use any protective clothing, although those handling munitions are in Level D PPE. A facility may be required for workers who prepare and repackage munitions to suit up and take off the PPE. Any contaminated PPE or other equipment is disposed of in the Dynasafe detonation chamber.

The technology vendor states that boxed munitions can be removed at any time from the loading chambers and that once in the detonation chamber, sufficient residual heat remains to destroy the munition, even if there is no external energy (i.e., electricity for the resistance heaters) to further heat the chamber.

The monitoring instrumentation used (e.g., MINICAMS), location of the monitors, and monitoring procedures to be followed if Dynasafe were used in the United States for destroying non-stockpile chemical materiel are to be determined. Minimal agent monitoring equipment is used with the Dynasafe at the GEKA facility, as a result of an operating philosophy that emphasizes robust engineering, vapor containment, and extensive offgas treatment.

Public and Regulatory Acceptability in a U.S. Context

Although Dynasafe has not been permitted for use in the United States for chemical munitions, it will be undergoing extensive operational use with German chemical munitions and will be required to meet all European Union environmental regulations. The Dynasafe manufacturer believes that it will also be able to meet all U.S. environmental regulations, although this remains to be demonstrated. If operated as a closed system, postdetonation gases can be held in the detonation chamber and monitored for agent. If any agent is detected, heating of the gases can be continued until agent concentration drops to an acceptable level before the gases are processed further. This ability to hold and test the gases prior to either continued heating in the chamber or release to offgas processing equipment should increase the acceptability of Dynasafe technology to U.S. regulators and interest groups. If operated as an open system, the offgases are further treated and any remaining agent is destroyed in an afterburner (combustion chamber). If this treatment is viewed as an incineration step, it may be considered to be a negative factor in terms of the acceptability to the public and to regulators.

Odors, vibrations, noises, and other sensory impacts should not be noticeable to the public while the Dynasafe static kiln is in operation. The detonation takes place in a thick, double-walled chamber inside a containment structure, and the external impacts, if any, should be minimal.

Secondary Waste Issues

As noted above, offgases can be cleaned, tested, and treated prior to release. The scrap metal removed from the bottom of the detonation chamber is claimed to meet GPL requirements. If the chemical munitions contain tarry agent heels from polymerized or thickened mustard agent, then it may be difficult to destroy this material in the detonation chamber. In that situation, prolonged postdetonation treatment via continued heating of the metal in the clean-burning period may be required. This clean-burning period is expected to last several hours, but the actual time required for the agent concentration to be reduced enough to meet the GPL is not known.

A second issue regarding waste treatment involves the accumulation and disposal of arsenic following the detonation. The technology proponent acknowledges arsenic will accumulate on the walls of the detonation chamber and states that the arsenic will be removed from the chamber walls by subsequent detonations and that the chamber can be steam cleaned to remove the arsenic. Removal of arsenicals in the offgas is also an issue. Since arsenic will be present in some of the munitions to be destroyed, e.g., Clark-type agents in the German grenades, its treatment and recovery will take place in an ionizing wet scrubber to remove arsenic-containing dust.

The Dynasafe technology generates some liquid wastes. These come from the use of steam to clean the detonation chamber, from the quench tank, and from various scrubbers used to treat the offgas. The volumes are small compared with those generated from agent neutralization technologies.

Process Cost Issues

Although no quantitative cost information was available to the committee, qualitative factors indicate that the Dynasafe SK2000 static kiln could be cost effective when used to destroy chemical munitions that are commensurate with its size. The Dynasafe SK series of static kilns is a well-established product line routinely used to destroy conventional explosively configured small arms and munitions. Thus, there is an operational track record to indicate that they can compete with other methods for destroying such items. One version of the Dynasafe kiln is being used by the NSCMP to destroy bursters in a burster detonation vessel at the Munitions Assessment and Processing System facility in Edgewood, Maryland. The acquisition cost of this unit should provide a benchmark for estimating a comparable cost for a Dynasafe unit used for chemical munition processing since the operation of the loading and detonation chambers should be similar.

As of the preparation of this report, the Dynasafe static kiln had been used to destroy some German chemical weapons; however, cost data for operating the kiln were not available. Since the kiln only requires two staff to operate and two to four more for supervision and in a control room, labor costs are expected to be low. A more substantial cost component may be for operating and maintaining the fairly complex offgas treatment system (e.g., a cyclone, a combustion chamber, quench, scrubbers, and filters) used in conjunction with the Dynasafe static kiln when processing chemical munitions. The complexity of the gas treatment system will depend on the offgas constituents to be treated, regulatory requirements, and whether or not the system is operated in a continuous (open) or batch (closed) mode. Thus, it is not possible to estimate the capital and operating costs for a Dynasafe offgas treatment system in the United States based on the experience in Germany, although the complexity of that system may suggest an upper bound on such costs.

As with other munition destruction systems, the Dynasafe will incur costs for setup, teardown, regulatory compliance, monitoring, lab support, and disposal of treated residuals such as metal fragments. The magnitude of these and other operating costs will depend on the specific application, the duration of operation, state and federal permit requirements, and the nature of the materiel to be treated.

Summary

The Dynasafe technology has been demonstrated to be effective in destroying small conventional munitions and explosives, in destroying some chemical agents, and in destroying mustard agent-filled, explosively configured German grenades. If, during continued operation at GEKA in destroying German munitions containing a variety of agent fills (which was in progress as this report was being prepared), the Dynasafe static kiln demonstrates the ability to safely and effectively access the agent in such munitions, destroy the chemical agents inside, and process secondary wastes, then it could be a viable technology for use in disposing of U.S. non-stockpile chemical munitions.

The Dynasafe technology could find application at U.S. sites where fairly large numbers of chemical munitions such as bomblets, mines, 105-mm projectiles, and 155-mm projectiles are recovered and where effective use could be made of its high throughput capacity. Its limited explosive containment capacity, however, limits it to destroying items of up to 5 pounds TNT-equivalent, about the same as the EDS-2. This limited capacity also places a requirement on the Dynasafe operator to not introduce high-explosive rounds into the Dynasafe detonation chamber that would exceed the chamber's explosive containment capacity. Even with a 100 percent safety margin—allowing up to 10 pounds TNT-equivalent of explosive loading—the detonation of such rounds could reduce the life of the chamber and, as a worst case, could severely damage it.

The Dynasafe technology depends on heat rather than donor charges to destroy energetics within a munition and to access the agent fill. This process is expected to be effective for chemical munitions that contain energetics but may be more problematic for inert chemical munitions if the munition emerges from the detonation chamber intact and in situ agent destruction needs to be confirmed. Such confirmation will be required to verify agent destruction does take place. Following this verification of agent destruction, the Dynasafe static kiln can be considered to be an effective and flexible technology for destroying large quantities of chemical munitions within its explosive containment and munition size constraints.

COMPARATIVE EVALUATIONS OF TIER 1 MUNITIONS PROCESSING TECHNOLOGIES

As defined in detail in Chapter 3, the committee used five basic evaluation factors to assess the status of Tier 1 technologies. These factors were commented on earlier in this chapter in the respective evaluation factors analysis sections for each of the three Tier 1 international munitions processing technologies.

TABLE 4-9 Evaluation Factor Rating Comparison of Tier 1 Munitions Processing Technologies with U.S. EDS

	Evaluation Factors (Rating[a])[b]				
Technology	Process Maturity	Process Efficacy/ Throughput	Process Safety	Public and Regulatory Acceptability in a U.S. Context	Secondary Waste Issues
U.S. EDS	+	+	+	+	0
CDC	+	+	+	0	0
DAVINCH	+	+	+	0[c]	+
Dynasafe	+	+[d]	+	0	0

[a]Legend: +, acceptable; 0, partially acceptable; –, unacceptable; ?, inadequate information.

[b]Costs associated with purchasing and operating a given technology would also be a significant criterion, but the committee did not have access to capital or operating cost data.

[c]DAVINCH is more likely to be acceptable to the public than the CDC and Dynasafe because of its demonstrated ability to hold and test waste gases, but it has not yet been permitted (see the section "Public and Regulatory Acceptability in a U.S. Context" in Chapter 4).

[d]Rating is contingent on the ability of the Dynasafe process control system to confirm agent destruction in all munitions that do contain agent.

Table 4-9 rates the Tier 1 munitions processing technologies according to these evaluation factors and compares them to the EDS technology that is presently in use by the NSCMP. The symbols used in the ratings scheme are also defined in more detail in Chapter 3.

The committee next considered several engineering parameters important to any comparison of these technologies. This comparison is presented in Table 4-10 for specific versions of each of the technologies rated in Table 4-9. The importance of these engineering parameters can be indicated as follows:

- *Throughput rate*. Maximum throughput rate may not be important for the disposal of small numbers of munitions but may be significant where a large number of munitions are to be destroyed. The estimated daily throughput rates for the three detonation technologies are compared in a more quantitative fashion in Table 4-11.
- *Destruction verification capability*. Whether the agent destruction can be confirmed before the liquid or gas is released to secondary treatment (hydrolysate disposal or offgas treatment) may be a consideration that is important to public stakeholders and regulators. This is often referred to as a hold-test-release capability.
- *Largest munition*. The largest munition and the largest explosive loading that can be handled by a specific unit will be important in assessing which technologies should be considered for a given mix of munitions.
- *Reliability/operability*. The experience that a given type of system has accumulated in processing conventional and chemical munitions is a significant factor indicator in the choice of technology.
- *Transportability*. Whether a specific technology is transportable—that is, whether it is movable from place to place, as required, or must be built as a fixed facility—may be a significant factor in selecting a technology for a given or anticipated scope of work (number and sizes of munitions, agent types, etc.) at a specific location.

Chamber lifetime is among the considerations that would have a significant impact on cost, reliability, and safety. Were the U.S. Army to further investigate any of the detonation-type technologies examined in this report, a structural integrity assessment for the number of detonation cycles that could be anticipated for the life of the detonation chamber with respect to the types of munitions to be processed would give important information. Likewise, a failure modes and effects analysis for each type of detonation system under consideration would be highly desirable.

The American Society of Mechanical Engineers (ASME) has formulated design codes to ensure the safe and reliable operation of pressure vessels. ASME has formed a committee to examine the design of pressure vessels subjected to intermittent impact loadings (i.e., vessels in detonation services). Two of the companies that supply detonation chambers (DAVINCH and CDC) have representatives on that committee. The committee responsible for this report understands that the design requirements for pressure vessels subjected to intermittent impact loadings will be defined in a Code Case that is essentially an addendum to the ASME Section VIII pressure vessel code. The ASME Code includes significant safety factors in terms of the yield and ultimate strength values that are used and, where appropriate, requirements for impact testing. In reply to specific questions, each of the suppliers of detonation chambers indicated that they will be able to comply with the requirements of the ASME Code for pressure vessels subjected to intermittent impact loadings.

In general, costs associated with purchasing and operating a given technology constitute a significant criterion, but the committee did not have access to data on capital or operating costs. Similarly, when considering a technology choice, the composition, or anticipated composition of the munitions to

TABLE 4-10 Specific Engineering Parameters for Existing Munitions Processing Technologies

Technology Model	Throughput Rate	Destruction Verification Capability	Largest Munition	Reliability/Operability	Transportability
EDS-2	1 detonation every other day; up to 6 munitions per detonation	Liquid and gaseous effluents can be held and tested before release	5 lb TNT-equivalent; wide range of weapons acceptance; maximum: 155-mm projectile; physical size of munition determines throughput rate	Extensive experience with chemical munitions	Fully transportable; 1 trailer
CDC (TC-60)	Up to 20 detonations per 10-hr shift; estimated potential throughput given by technology proponent as 22-40/day; actual will be determined in 2006	Monitoring of offgas prior to release to carbon adsorption bed system	60 lb TNT-equivalent; 210-mm projectile	Extensive experience with conventional munitions; has demonstrated reliability; 4 years experience in production mode without failure	Transportable on 8 tractor trailers
DAVINCH (DV-60)	Yellow bombs: 9/day Red bombs: 18/day 75-mm, 90-mm munitions: 36/day	Detonation gases held in tank and tested for agent before decision made to release or provide additional treatment	65 kg TNT-equivalent; expected to be an 8-in. projectile or a small bomb	Experience with destruction of 600 Japanese Red and Yellow chemical bombs containing various agents	DV-60 designed to be a fixed facility, not transportable
Dynasafe (SK2000)	Varies greatly with munition and operating mode; if used as an open system (continuous mode), sample throughput rates are 20/day for 8-in. projectile, 40/day for 155-mm projectile, 120/day for 105-mm projectile and 4.2 in. mortar round	Open system (continuous mode): none prior to offgas treatment; closed system (batch mode): hold and test in expansion tank	5 lb TNT-equivalent; 8-in. projectile, if fragment shield used to protect chamber; up to 750-lb bomb if most of agent is drained first	Extensive experience with conventional munitions; some experience with German chemical munitions	SK2000 designed to be a fixed facility, not transportable

TABLE 4-11 Estimated Daily Throughput Rates for Three Detonation Technologies (10-hr Day)

Munition	CDC TC-60	DAVINCH DV65	Dynasafe SK2000
4.2-in. mortar, M1	40	36	120
75-mm projectile, M64	40	30	270
5-in. projectile, MK VI	22	18	120
5-in. projectile, MK 54	22	12	90
155-mm projectile, MK II	22	12	40
8-in. projectile, T174	22	6	20
Bomblet, M139	60	72	480
105-mm projectile, M60	22	30	120
100-lb bomb, M47	6[a]	6	20[b]
115-lb bomb, M70	5[a]	6	20[b]

[a]Bomb is drained into 20-lb lots and each lot separately destroyed in CDC-60.
[b]Bulk of agent is removed before treatment in SK2000. Drained agent and the item are treated separately.

SOURCES: CDC: CH2MHILL response to committee questions of February 6, 2006; DAVINCH: information provided by Kobe Steel, Ltd., to the committee on March 25, 2006; Dynasafe: information provided by UXB International to the committee on February 15, 2006.

be destroyed would also be an important factor. Insofar as it is uncertain which non-stockpile sites may be chosen for remedial action in the near term (post 2007) future,[24] as is the amount of resources that would be dedicated to recovery operations and thus the rate of recovery, the committee did not address how a technology or mix of technologies might be implemented for a specific site situation. Moreover, there is considerable uncertainty surrounding the Army's site inventory data in terms of the specific conditions, relative locations, remaining amounts of agent fills, and other characteristics of munitions to be encountered during recovery operations.

The committee also addressed the subfactors given in Chapter 3 for each of the five main evaluation factors. The subfactor questions for the Tier 1 international technologies that are suitable for munitions processing along with the EDS technology are addressed in Tables B-1 through B-5. These tables provide a convenient side-by-side means for comparing some specific aspects of the technologies in terms of the available data and the expert judgment of the committee.

FINDINGS AND RECOMMENDATIONS

Finding 4-1. The U.S. Army's EDS, although proven to be safe and effective, has a low throughput rate, is limited in the size of the munitions it can handle, and generates a liquid waste stream that must be disposed of. Consequently, while it will continue to have application for small quantities of munitions, EDS would be expected to have limited applicability to the destruction of the anticipated large quantities and variety of munitions and agent-contaminated items expected to be found at large burial sites in the United States.

Finding 4-2. Detonation-type technologies offer complementary capabilities to the EDS and all have the following characteristics:

- There is no agent neutralization step.
- All are total solutions—that is, they all access the agent, destroy the energetics and agent, and decontaminate the munition bodies.
- All require secondary thermal or catalytic treatment of offgases.
- All have a higher throughput than the EDS and the same or greater explosive containment capability.
- All have been operated safely.

Finding 4-3. The CDC is a mature technology that has destroyed 2,500 chemical munitions in Belgium. Additional testing in the United Kingdom has pointed to its acceptability in the United States in terms of efficacy and safety. Public acceptance might be qualified because the CDC is not a hold-test-release system, although it has been used here for the destruction of conventional munitions. It is the only one of the three detonation-type technologies that at present can be considered to be transportable (but mobile versions of the other two types of detonation technologies have been designed).

Finding 4-4. Of the detonation-type technologies, the DAVINCH is the only one that currently has demonstrated the ability to hold, sample, and analyze waste gases prior to releasing them into the offgas treatment system. It has the largest explosive containment capacity of the detonation-based technologies and appears to be suitable for destroying moderately large quantities of a large variety of chemical munitions.

Finding 4-5. The Dynasafe static kiln technology has been demonstrated to be effective in destroying small conventional munitions and explosives, small chemical munitions containing explosives, and in destroying some chemical agents. The ability to confirm the release and destruction of agent contained in chemical munitions that do not contain energetics needs to be demonstrated. The Dynasafe technology appears to be suitable for destroying large quantities of small to medium-sized chemical munitions.

Finding 4-6. Each detonation-type technology has different characteristics such as destruction rate, initial capital and operating costs, and ability to be moved from one location to another that are relevant to the selection of a system for a particular project. Structural integrity, defined as a specified allowable number of detonation cycles, is another factor to be considered, as would be the results of any failure modes and effects analyses.

Recommendation 4-1. The U.S. Army should select a detonation-type technology for destroying recovered chemical munitions excavated from a large burial site, although the EDS will continue to have application, especially at small sites. In view of the rapidly evolving development efforts on the three international detonation-type technologies, the U.S. Army should monitor the operations and capabilities of these technologies and collect cost and performance data with the goal of selecting one of them as the primary technology.

Finding 4-7. Procedures for measuring the destruction and removal efficiency (DRE), destruction efficiency (DE), or some other metric of performance for detonation-type processes do not appear to have been established in the United States. This gap will seriously hinder future evaluations of such technologies for possible application to non-stockpile

[24] As noted in Chapter 2, following completion by April 29, 2007, of the Chemical Weapons Convention treaty requirements applying to CWM that has already been recovered, no specific subsequent site remediation mission had been defined for the NSCMP at the time this report was being prepared.

operations. Such destruction and removal information is important for both regulators and the public.

Recommendation 4-2. To further the evaluation of detonation-type technologies for non-stockpile applications, the U.S. Army should establish accepted procedures that effectively and efficiently determine the degree of agent destruction or in some other way measure the performance of these processes. The procedures should involve the feeding of complete munitions to the process—that is, munitions containing either agent or a chemical surrogate that is more difficult to destroy than the chemical agent that is most resistant to destruction. Both the degree of agent destruction in the actual detonation event and the degree of agent destruction in the system overall should be determined. Such procedures should be developed with input from all of the relevant stakeholders.

REFERENCES

Measurement of Performance

ATSDR (Agency for Toxic Substances and Disease Registry). 2005. Public Health Assessment for Oak Ridge Reservation (USDOE) TSCA Incinerator, U.S. Department of Energy Oak Ridge Reservation, December 27. Available online at <http://www.atsdr.cdc.gov/HAC/PHA/TSCA/FINAL_TSCA_PHA_2-2-06.pdf>. Last accessed March 10, 2006.

Controlled Detonation Chamber

Bixler, B. 2005. Controlled Detonation of Chemical Weapons. Available online at <http://www.dstl.gov.uk/news_events/conferences/cwd/2005/proceedings55.pdf>. Last accessed February 28, 2006.

Blades, T.A., R. DiBerardo, G. Misko, and N. McFarlane. 2004. Demonstration/Validation of the TC-25 Donovan Blast Chamber Porton Down, U.K., Final Demonstration Test Report, April-September 2003, ECBC-TR-362, May. Aberdeen Proving Ground, Md.: Edgewood Chemical and Biological Center, Chemical Biological Applications and Risk Reduction Business Unit.

De Bisschop, H.C., and T. Blades. 2002. Destruction of Chemical Weapons: Evaluation of the Donovan Contained Detonation Chamber (CDC) in Poelkapelle, Belgium. ECBC-SP-010, July. Aberdeen Proving Ground, Md.: Edgewood Chemical and Biological Center.

DeMil International. 2004. System Hazard Analysis: TC-60 Controlled Detonation Chamber, Draft Report, June 7. Aberdeen Proving Ground, Md.: U.S. Army Edgewood Chemical-Biological Command.

DeMil International. 2005a. Throughput Analysis Controlled Detonation Chamber System, March. Huntsville, Ala.: U.S. Army Engineering Support Center.

DeMil International. 2005b. Deployment Plan Controlled Detonation Chamber System, March. Huntsville, Ala.: U.S. Army Engineering Support Center.

DiBerardo, R. 2004. Demonstration/Validation Testing of the Controlled Detonation Chamber (CDC). Available online at <http://www.dstl.gov.uk/news_events/conferences/cwd/2004/proceedings21.pdf>. Last accessed March 1, 2006.

NRC. 2002. Systems and Technologies for the Treatment of Non-Stockpile Chemical Materiel. Washington, D.C.: National Academy Press.

NRC. 2005. Impact of Revised Airborne Exposure Limits on Non-Stockpile Chemical Materiel Program Activities. Washington, D.C.: The National Academies Press.

U.S. Army. 2001. Memorandum for Record: Initial Assessment of the Donovan Controlled Chamber (CDC) Used in Belgium from 14 May through 22 June, June 28. Aberdeen Proving Ground, Md.: Program Manager for Chemical Demilitarization.

DAVINCH Cold Detonation Chamber

Asahina, J. 2004. Kanda Project, Its Outline and Public Acceptance. Available online at <http://www.dstl.gov.uk/news_events/conferences/cwd/2004/proceedings36.pdf>. Last accessed March 1, 2006.

Asahina, J. 2005. Destruction of OCW at Kanda Project. Available online at <http://www.dstl.gov.uk/news_events/conferences/cwd/2005/proceedings28.pdf>. Last accessed March 1, 2006.

Asahina, J., K. Koide, and K. Kurose. 2005. "DAVINCH" Controlled Detonation Process Applied to Destroy 50 kg Yellow Bombs and 15 kg Red bombs at Kanda. Available online at <http://www.dstl.gov.uk/news_events/conferences/cwd/2005/proceedings47.pdf>. Last accessed March 1, 2006.

Lefebvre, M.H., S. Fujiwara, and J. Asahina. 2005a. Disposal of Old Chemical Weapons by Controlled Detonation: Performance Analysis of the Kanda Port Campaign. Available online at <http://www.dstl.gov.uk/news_events/conferences/cwd/2005/proceedings15.pdf>. Last accessed March 1, 2006.

Lefebvre, M., S. Fujiwara, and J. Asahina. 2005b. "Disposal of non-stockpile chemical weapons by controlled detonation." Theory and Practice of Energetic Materials, Vol. 6, Proceedings of the 2005 International Autumn Seminar on Propellants, Explosives and Pyrotechnics. W. Yajun, H. Ping, and L. Shengcai, eds. Beijing, China: State Key Laboratory of Explosion Science and Technology.

Dynasafe Static Detonation Chamber

Dynasafe. 2006. Mobile Demilitarization System for Munitions: Static Kiln SK1200CM and Off-Gas Cleaning, January. Karlskoga, Sweden: Dynasafe AB.

Ohlson, J., H. Weigel, T. Stock, H. Tsuboi, and K. Yokoyama. 2004. Destruction of CW Type Ammunition Shells Filled with Surrogate Agents in a DYNASAFE Static Kiln SK2000. Available online at <http://www.dstl.gov.uk/news_events/conferences/cwd/2004/proceedings10.pdf>. Last accessed March 2, 2006.

Weigel, H., J. Ohlson, and T. Stock. 2004. The DYNASAFE Static Kiln SK2000: Its Application for Old Chemical Munitions Destruction at Munster. Available online at <http://www.dstl.gov.uk/news_events/conferences/cwd/2004/proceedings08.pdf>. Last accessed March 2, 2006.

5

Tier 1 International Agent-Only Processing Technologies

INTRODUCTION

In addition to the detonation-type technologies for munitions processing described in Chapter 4, the committee recognized that certain other types of processes could be used to destroy chemical agents, if not entire chemical munitions. The Non-Stockpile Chemical Materiel Project (NSCMP) already employs equipment that is based on one of these agent-only technologies. The rapid response system (RRS) and the single container agent neutralization system, described in Chapter 1, are good examples. These systems do not destroy complete munitions but use neutralization to destroy small amounts of agent contained in chemical agent identification sets. Several other technologies for agent-only destruction were identified and investigated. Of these, two—(1) the Russian two-stage process of neutralization with addition of bitumen and (2) incineration—were designated as Tier 1 technologies for agent-only destruction. These technologies were investigated more fully, and are described and evaluated in this chapter.

The Russian two-stage process employs neutralization (hydrolysis) as a first stage. In a second stage, the product of the neutralization step is added to bitumen, yielding a stable bitumen-salt product that can be landfilled or stored long term. In the United States, neither the non-stockpile program nor the stockpile program for chemical demilitarization employ technology similar to the Russian two-stage process.

On the other hand, incineration is employed on a large scale in the stockpile program. The U.S. Army's baseline incineration system includes a process for reverse assembly (disassembly) of the stored munitions and collection of the contained agent. As such, it constitutes a complete munition processing system for munitions that have been properly stored. Agent drained from these munitions is sent to a liquid incinerator for destruction, while energetics and munition bodies are sent to separate furnaces for destruction or decontamination. This baseline system is now in use at four stockpile sites in the United States, where a variety of obsolete munition types containing nerve and mustard agents have been stored.

However, because of the poor condition of recovered non-stockpile items in general, it is not practical to employ a reverse assembly process for the munitions dealt with in non-stockpile operations. On the other hand, the agent destruction part of the baseline incineration process is applicable to recovered bulk agent or to agent that has been recovered from munitions. Three European countries currently employ incineration to destroy agent recovered from old and abandoned chemical weapons. However, the agent recovery methods used are considered out of date, and their future use is not recommended.

The committee designated additional technologies that destroy only agent, including acid digestion, biological treatment, the Defence Science and Technology Laboratory (DSTL) electric furnace, electrochemical treatment, photocatalysis, and plasma arc technology, as Tier 2 technologies. They were investigated to a lesser extent and are covered in Chapter 6.

In this chapter, the committee briefly reviews some other applications of neutralization and hydrolysis elsewhere in the world to provide historical perspective for the technology being used in the United States to destroy recovered chemical weapons, as described in Chapter 1. Detailed descriptions and evaluations of the Russian two-stage neutralization/bitumenization process and incineration processes follow that review.

USE OF NEUTRALIZATION AND HYDROLYSIS IN THE REST OF THE WORLD

Neutralization and hydrolysis[1] are accepted and widely used technologies that have been used to treat a variety

[1]As previously noted in Chapter 1, this report employs "neutralization" for the use of an organic reagent to destroy agent and "hydrolysis" for the

of chemical agents. Their use in the United States is summarized in Chapter 1. Other countries have also employed neutralization and hydrolysis. This brief review examines some instances where this technology has been used and describes some of the technical variations in how it has been applied.

Hydrolysis was used for the destruction of agent production facilities and chemical weapons in Iraq from 1991 to 1994 (Manley, 1997). Iraq had various weapons containing the nerve agents GA, GB, and a mixture of GF (O-cyclohexyl methylphosphonofluoridate) and GB. The latter mixture also contained up to 40 percent dichloromethane. First, all energetics were removed from the weapons. Several techniques were then employed to remove agent from the weapons, including removal of the filling plug when possible, drilling holes, and using shaped explosive charges to create holes. The drained weapons were rinsed, again using various techniques, and then burned. A former nerve agent production plant was converted into a hydrolysis facility, with a glass-lined, 3 cubic meter reactor used to carry out hydrolysis in batchwise fashion. A 15 weight percent solution of aqueous sodium hydroxide was used to treat 300-350 liters of agent per batch. Use of higher concentration caustic resulted in blockages when emptying the reactor. The reaction was considered complete when the agent concentration dropped below the detection limit of 1 ppm and the caustic concentration remained constant for two successive measurements. The hydrolysis products were pumped to a concrete-lined lagoon and allowed to evaporate to dryness. The dry residue was mixed with concrete and entombed in more concrete. The plant capacity was 1-1.5 metric tons per day. About 70 metric tons of GB and GB/GF mixture were destroyed. The same plant was used to destroy the GA recovered from 75 metric tons of impure product. The resulting hydrolysis product contained sodium cyanide and was placed in steel tanks that were sealed into large concrete storage bunkers.

Pearson and Magee (2002) state that neutralization is used at the Belgian chemical weapon dismantlement facility at Poelkapelle to destroy some of the approximately 60 combinations of agents used in World War I. The hydrolysis products are sent to the INDAVER commercial hazardous waste incinerator near Antwerp. Mustard agent is neutralized using an aqueous ethanolic caustic solution. Diphosgene, which sometimes contains phosgene in solution, and chloropicrin are treated with aqueous sodium hydroxide solution. Other agents are dissolved in butanol and incinerated without hydrolysis, although neutralization techniques for these were being investigated.

use of an aqueous reagent to destroy agent. Neutralization and hydrolysis are often used interchangeably in the literature on chemical agent demilitarization. Hydrolysis is the more appropriate term from a chemical process perspective. Neutralization is more in keeping with the notion of "to neutralize and thereby render innocuous." It may be found in the literature to refer to hydrolysis in aqueous or nonaqueous media.

Hydrolysis with aqueous sodium hydroxide was used to destroy mustard agent, lewisite, and a mixture of mustard agent and lewisite contained in unearthed beer bottles in Samukawa, Japan (Kawataki et al., 2005). Several hundred bottles were involved. The tops were removed from the bottles, and the agent was removed and filtered. The agent was then fed to a stirred reactor containing caustic. The product from this first reactor was fed to a second reactor that contained an unspecified oxidation reagent. The product from the second reactor was analyzed to ensure that agent concentrations were below the threshold value of 0.05 mg/L for HD and 0.1 mg/L for lewisite. The liquid effluent was then passed to an industrial waste contractor; the ultimate fate was not reported. Some bottles contained solid wastes. These were treated to a U.S. 5X standard for decontamination by heating in an electric furnace.[2] The electric furnace was also used to decontaminate the empty bottles and other solid waste to a 5X condition. Exhaust gases from both operations were treated before discharge using caustic scrubbing and adsorption on activated carbon.

A Russian procedure for destroying mustard agent/lewisite mixtures and lewisite itself by caustic hydrolysis appears to be similar to that used at Samukawa. According to a summary in *RIA Novosti*, the destruction of the 241 tons of lewisite stored at Gorny was expected to be completed in December 2005 (Litovkin, 2005). A lewisite neutralization facility being built with German assistance at Kambarka was to go online in December 2005. There are 6,349 tons of lewisite stored there. The hydrolysate will be incinerated (GSN, 2005). It appears that the Gorny facility served as a pilot plant to demonstrate the technology.

Over 700 tons of mustard agent were destroyed by hydrolysis in Canada (Sutherland, 1997). The destruction was carried out at the Defence Research Establishment Suffield (DRES) in 1975-1976. Alkaline hydrolysis was employed and the products were incinerated. From 1989 to 1991, nerve agents (G and V classes) were removed from weapons using drill-and-drain methods and neutralized using 20 percent potassium hydroxide in methanol. The neutralization products were then incinerated. Lewisite was neutralized with an alkaline peroxide solution containing sodium or calcium chloride. The products included sodium or calcium arsenate, acetylene, and chloride salts.

[2]The term "5X" until recently was used by the U.S. Army to indicate that an item had been decontaminated completely of the indicated agent and could be released for general use or sold to the general public in accordance with applicable federal, state, and local regulations. The approved method for establishing a 5X condition was to heat the item throughout to 538°C (1000°F) for 15 minutes. This was considered sufficient to destroy chemical agent molecules. With the promulgation of revised airborne exposure limits (AELs) by the Centers for Disease Control in 2003/2004, the Army has moved to these new AELs for purposes of material and waste classification and has been substantially revising its entire X classification system accordingly.

The article by Sutherland also mentions the use in Russia of a combination of neutralization and incineration. The nerve agents GB and GD, and also mustard agent, were reacted with ethanolamine, and the products of these reactions were subsequently incinerated. VX was reacted with a 1:1 mixture of ethylene glycol and orthophosphoric acid, again with the products sent for incineration. Some 300,000 kg of toxic substances were destroyed in a facility that was too small to be considered industrial-scale.

A neutralization process using hydrogen peroxide to destroy agent is one of the technologies currently being used in Civitavecchia, Italy, at the NBC Joint Technical Logistics Center, which is known by its acronym, Ce.T.L.I. NBC. Demilitarization activities are focused on stockpiled munitions containing primarily Clark I agent but also a mixture of mustard agent and Clark I, chloropicrin, or adamsite (DM).[3]

An automated plant that went into full operation in 2003 has destroyed thousands of rounds ranging in size from 65 mm to 155 mm and larger. Rounds are identified using high-energy x-ray analyses (by means of a linear accelerator-type x-ray source capable of penetrating 10 cm of steel). The liquid agents are frozen in the bottom of the round by chilling the round in an upright position. The round is opened by unscrewing, drilling, or cutting; after the agent is liquefied by warming to ambient temperature, it is emptied into a container for neutralization. The emptied round is cleaned with concentrated nitric acid and bleach. The plant destroys 64 projectiles in an 8-hour work day and uses antiaerosol and activated carbon filters to mitigate air emissions.

The Clark I and the blister agent mixtures are destroyed using hydrogen peroxide, producing an acidic, arsenic-containing solution, which is neutralized with lime. In contrast, adamsite (in the form of hard, solid cylinders) is milled to form a powder. The products of blister agent neutralization and adamsite milling are mixed with sand, cement, and water. The mixtures are placed in containers made of reinforced concrete, and the resulting blocks are stored outside on platforms. The rainwater that collects below is checked periodically for the presence of arsenic.

Munitions recovered from excavation or retrieved from the sea are also destroyed at Civitavecchia if they are judged to have sufficient integrity to be safely transported. Badly damaged rounds that cannot be transported to Ce.T.L.I. NBC are handled on-site. A render-safe procedure is performed with a small shaped charge. Upon detonation, the shaped charge removes the fuze without initiating the burster and also opens an access hole for the agent fill. Technicians manually remove the agent from the munition and transport it to Civitavecchia for neutralization.

[3]Personal communication between LTC Charles Davis, Assistant Army Attaché, U.S. Defense Attaché Office, Rome, Italy, and Gary Groenewold, committee member, March 28, 2006.

RUSSIAN TWO-STAGE PROCESS: NEUTRALIZATION WITH ADDITION OF BITUMEN

Description

The focus in Russia in recent years has been on a significantly different approach that involves neutralization and addition of the neutralent to bitumen to form a stabilized mass that can be safely stored in drums in a vault for some long period of time or landfilled.[4] That approach is the focus of this section.

The chemical procedures involved have been developed for mustard agent, VX, VR (the Russian version of VX), GB, and GD. This technology was developed at the GOSNIIOCHT research institute in Russia (Krotovich, 1998). The procedures for VR, VX, GB, and GD were evaluated by a joint Russian/U.S. team (Bechtel, 1996).[5] The joint evaluation involved experimental studies carried out at the Saratov Higher Military Engineering School of Chemical Defense in Russia and at the U.S. Chemical-Biological Defense Command in Edgewood, Maryland. At both sites, Russian and U.S. scientists worked together to carry out the chemical procedures involved, analyze the products of the reactions, and measure the toxicity of the final product.

VR is neutralized in Stage 1 of the process with a decontamination solution known as RD-4 (Pearson and Magee, 2002). This solution contains potassium isobutylate dissolved in isobutanol and N-methylpyrrolidinone. The VR and the RD-4 are reacted at atmospheric pressure at 90°C-95°C for 30 minutes to break the P-S bond, which causes the high toxicity of VR, and form diisobutyl methyl phosphonate and other products (see Figure 5-1).

In the second stage of the process, the reaction mass is added to bitumen at a starting temperature of 130°C-140°C that is subsequently increased to 180°C under reduced pressure. This results in the formation of low volatility, low toxicity phosphorus-containing compounds and the vaporization of isobutanol and N-methylpyrrolidinone. The vapor stream is condensed to form a distillate that contains isobutanol and N-methylpyrrolidinone. Upon cooling, a solid bitumen-salt end product is formed. The joint evalu-

[4]Again, for clarity, the product of the reaction of an aqueous reagent with an agent will be referred to as a hydrolysate in this report. The product of a nonaqueous reagent with an agent will be referred to as a neutralent.

[5]Organizations participating in the Bechtel joint evaluation in 1996 were the Office of the Secretary of Defense, Cooperative Threat Reduction Program Office, U.S. Department of Defense; the U.S. Army Program Manager for Chemical Demilitarization; the U.S. Army Chemical and Biological Defense Command (Edgewood Research, Development, and Engineering Center); the On-Site Inspections Agency Interagency Affairs, Chemical Weapons Branch, U.S. Department of Defense; the Russian Federation Ministry of Defense; President's Committee for Conventional Problems of Chemical and Biological Weapons, Russian Federation; Saratov Higher Military Engineering School of Chemical Defense, Russian Federation; and the State Scientific Research Institute for Organic Chemistry and Technology, Russian Federation.

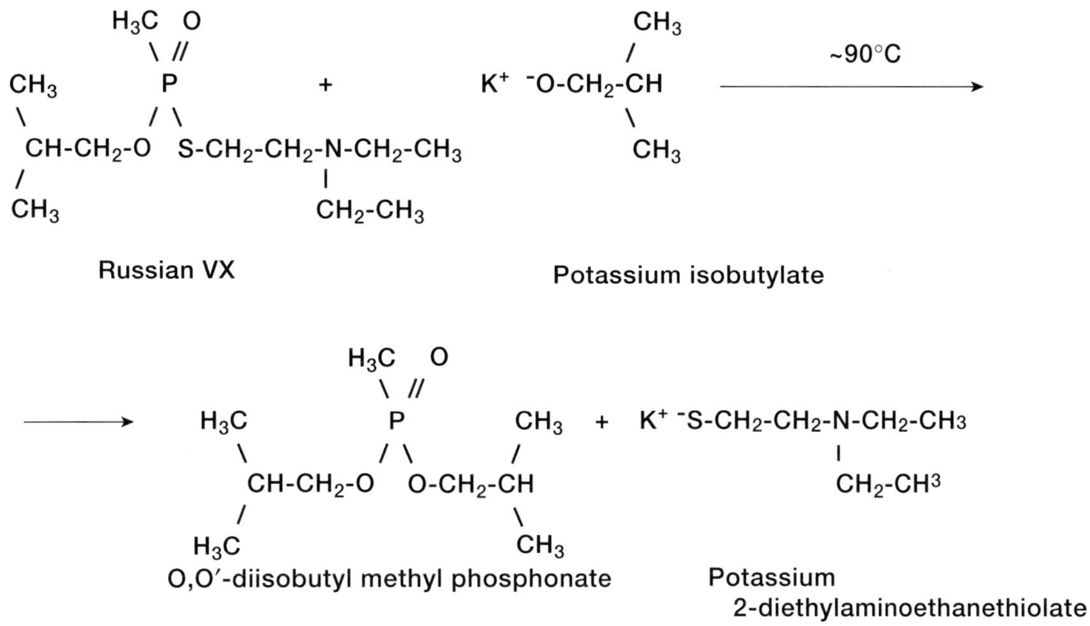

FIGURE 5-1 Reaction of Russian VX and potassium isobutylate. SOURCE: Bechtel, 1996.

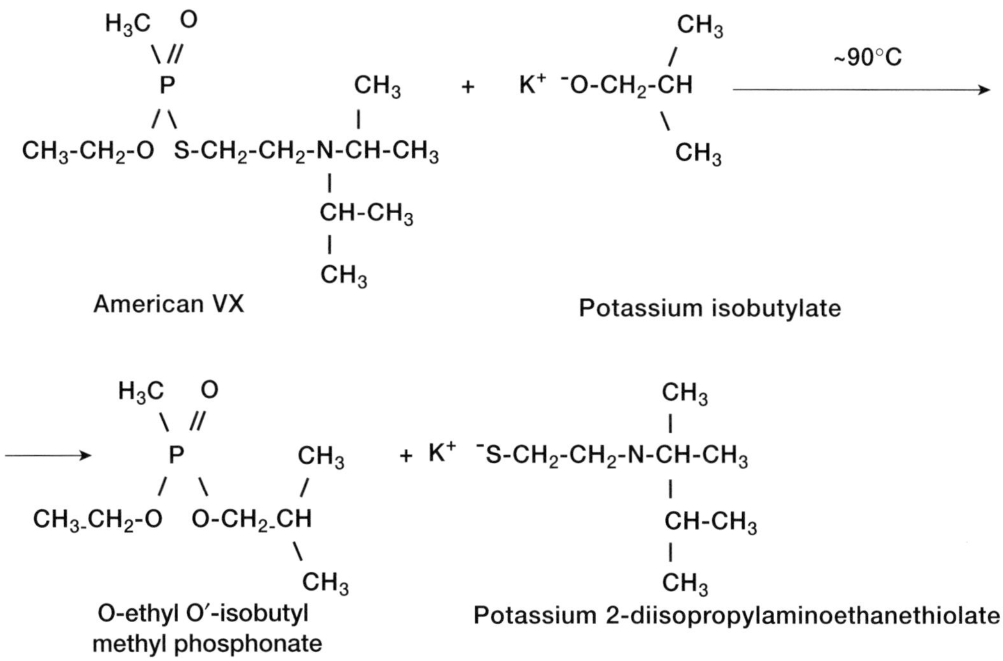

FIGURE 5-2 Reaction of VX and potassium isobutylate. SOURCE: Bechtel, 1996.

ation report does not specifically indicate the fate of the diisobutyl methyl phosphonate in Stage 2 but does state that the phosphorus-containing products are tightly bound in the bitumen and would be impossible (or at the very least extremely difficult) to remove, thereby satisfying the CWC requirement for irreversible destruction.

VX is also reacted with RD-4 under similar conditions, again rupturing the P-S bond. The stage 1 reaction is shown in Figure 5-2. The Stage 2 bitumen addition is then carried out in the same fashion as for VR.

The destruction efficiencies (DEs) reported by Flamm and Pakhomov were not defined but were given as ≥99.9 percent

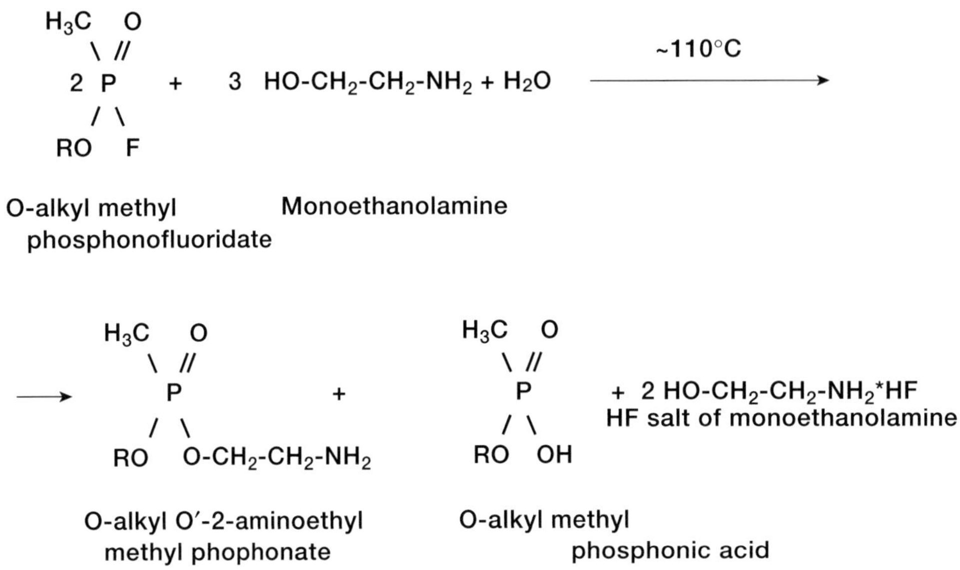

FIGURE 5-3 Notional reaction scheme for the addition of G-type agent to aqueous monoethanolamine (MEA). SOURCE: Bechtel, 1996.

for VR and VX for Stage 1 only and >99.999 percent for the overall process (Bechtel, 1996). VR and VX concentrations in the bitumen-salt end product were consistently below the gas chromatography detection limit of 1×10^{-4} mg/ml.

The joint evaluation report gives the results of toxicity testing on the final bitumen-salt products from the destruction of VR and VX. The measurement involved intragastric administration to rats, and the resulting LD_{50} was $>5 \times 10^3$ mg/kg. This is a very low toxicity, placing the material in the Russian State category "slightly dangerous," the same category as that of hexane, the solvent used in preparing the material administered to the rats.

About 3 kg of bitumen-salt product is produced for each kilogram of VX destroyed. About 1 kg of distillate is produced for each 2 kg of bitumen-salt product. The toxicity of the distillate was not measured.

Flamm and Pakhomov also describe the destruction of GB (sarin) and GD (soman) (Bechtel, 1996). The GB or GD is dissolved in industrial-grade ethanolamine containing about 20 percent water. The mixture is heated to 110°C for 1 hour at atmospheric pressure (Figure 5-3). The reaction thus breaks the P-F bond, which is what causes the high toxicity of the G compounds. The products from this reaction are added to calcium hydroxide and bitumen, then heated to about 200°C under reduced pressure for 1 hour. This treatment is expected to remove any last traces of GB and GD. Alcohols produced in the reactions and excess ethanolamine are distilled from the bitumenization mixture. The bitumen-salt mixture is drained from the reactor and allowed to solidify. About 5-7 kg of bitumen-salt product is produced for each kilogram of GB or GD destroyed. About 1 kg of distillate is produced for each 3 kg of bitumen-salt product. The toxicity of the distillate was not measured.

Again, the DEs reported by Flamm and Pakhomov were not defined but were given as ≥99.9 percent for GB and GD for Stage 1 only and >99.999 percent for the overall process (Bechtel, 1996). The GB and GD concentrations in the bitumen-salt product were generally below the gas chromatography limit of detection of 1×10^{-4} mg/ml.

The joint evaluation report gives the results of toxicity testing on the final bitumen-salt products from the destruction of GB and GD. The toxicity was again measured in rats, with LD_{50} results of $>5 \times 10^3$ mg/kg. This is a low level of toxicity and puts the material in the Russian State category Level III, which makes it as toxic as the monoethanolamine solvent.

The Russian and U.S. scientists involved in the joint evaluation project concluded that destruction efficiencies were acceptably high, the reactions involved were irreversible, and the toxicity of the final product, the bitumen-salt product, was acceptably low. They also concluded that the distillates produced contained materials that could be recovered or recycled, thus reducing their contribution to overall waste production (Bechtel, 1996).

Although not presented in the joint evaluation report, reaction conditions have also been developed for the destruction of mustard agent. Mustard agent is preferably destroyed by neutralization with an equal weight of a reagent that is a 9:1 mixture of monoethanolamine and ethylene glycol (Pearson and Magee, 2002). The reaction proceeds for 1 hour at 100°C-110°C. The reaction mass is added to a "tough road bitumen" having a softening temperature of 65°C-70°C, with calcium hydroxide added. The mixture is held at 170°C for 45 minutes. This produces a bitumen mass with a softening temperature of 81.5°C, enabling the mass to be formed into monolithic blocks that can be buried. Slightly different conditions are given by Krotovich (1998).

Country-by-Country Experience

Russia

A facility that will use neutralization and addition of the neutralent to bitumen technology is being built at Shchuch'ye in Russia (Ryan and McNelly, 2003). Construction was begun in March 2003. This facility will be used for destruction of much of the nerve agent now stored in Russia. Nerve agent (organophosphorous compounds) is said to account for 79.2 percent of Russia's total chemical agent stockpile of about 40,000 tons (Petrov and Trubachev, 2000). Weapons and Stage 1 product will be shipped to Shchuch'ye from other sites. The plant is being constructed in stages. A pilot facility with a capacity of 500 metric tons per year will be built first. This facility will then be expanded to its full capacity of 1,200 metric tons per year.

The Russians had once planned to use the monoethanolamine and ethylene glycol process for destruction of their stockpile of 622 tons of mustard agent stored at Gorny (Saratov Oblast). They had planned to operate two process lines at Gorny: one for straight mustard agent using monoethanolamine and ethylene glycol, the other using aqueous caustic to destroy lewisite and mustard agent/lewisite mixtures (see above). To save money and to meet the CWC 1 percent destruction deadline, they used the lewisite line to destroy the mustard agent (Ember, 2003). The project was completed on November 14, 2003. They are now storing the 1,260 tons of hydrolysate in drums. It appears that the two-stage monoethanolamine/ethylene glycol neutralization followed by bitumenization will not be used in Russia for mustard agent because the country apparently has no more mustard agent apart from mustard agent/lewisite mixtures. Russia does not intend to employ the technology for destruction of arsenic-containing agents.

Evaluation Factors Analysis

Process Maturity

As noted above, the mustard agent version of the process of neutralization and addition of the neutralent to bitumen will not be used in Russia on a production scale. The nerve agent versions of the technology have not been used on a large scale but will be in the future. All three versions were tested extensively at the GOSNIIOCHT research institute.

Process Efficacy/Throughput

The process appears to completely destroy nerve and mustard agents. The Russians do not plan to apply the technology to arsenic-containing agents. As indicated above, they obtained nearly complete destruction of agent, generally down to a limit of detection of 1×10^{-4} mg/ml and a DE of >99.999 percent.

Process Safety

Pearson and Magee (2002) point out that the first step, the neutralization, is carried out at about 100°C under atmospheric pressure and using monoethanolamine or the RD-4 solution; this step does not present any unusual safety hazards. The second step, bitumenization, is carried out at higher, but still modest, temperatures and under reduced pressure; it also presents no unusual safety hazards.

Public and Regulatory Acceptability in a U.S. Context

Neutralization processes to treat CWM have routinely achieved public and regulatory acceptability in the United States. As described in Chapter 1, the U.S. NSCMP's explosive destruction system (EDS), RRS, and single container agent neutralization system employ neutralization to destroy recovered agent. Large-scale neutralization projects for stockpiled agent have been or will be carried out at the Newport, Pueblo, Aberdeen, and Blue Grass sites. While the method of disposing of the neutralent/hydrolysate can be a significant issue, the use of neutralization as the first step of the agent destruction project is generally readily accepted.

Secondary Waste Issues

A single significant waste, the bitumen-salt product, is produced by the Russian two-stage process. For VR destruction, the volume is about three times that of the starting VR. For GB and GD, the volume is four to seven times that of the starting agent. In both cases, this waste contains methylphosphonate products, which might leach from the waste over time. This possibility might impact disposal decisions. Petrov and Trubachev (2000) express concern about the possible eventual degradation of the bitumen-salt product and mention alternatives to the bitumenization step. As indicated above, the bitumen-salt products are not very toxic. No information was found on leaching as measured, for example, by the toxicity characteristic leaching procedure test that is used for compliance with RCRA regulations, and such testing would be needed before the product could be landfilled in the United States. Radilov (2004) presents toxicity data and then recommends that certain "sociohygienic" monitoring be done near destruction facilities in Russia. The bitumen-salt product will apparently be placed in permanent storage, although the literature is not entirely clear on this subject.

An earlier report (Khripunov and Parshall, 1999) indicates that the product will be placed in steel drums in aboveground vaults. A later reference (Ember, 2005) states only that the product will be placed in concrete vaults (polygons). The area is marshy, and the integrity of the storage facility is a concern. Clarification of the ultimate disposition of the bitumen-salt product is needed.

Process Costs

Some cost information for the Shchuch'ye facility is provided by Ryan and McNelly (2003). The United States has committed to funding the facility. The expected cost of construction, systemization, training, and start-up is given as $888 million. This cost appears to be very modest for a major facility. The area around Shchuch'ye is poor and lacking in infrastructure, such as regional roadways, water supply, gas supply, a hospital, and communications facilities.[6] Russia and other countries, but not the United States, are funding the construction of this infrastructure. That cost is not included in the $888 million figure.

INCINERATION

Description

Incineration is the high-temperature oxidation of organic compounds. The carbon and hydrogen in the compounds are ideally converted completely to carbon dioxide and water. The chemical agents at issue here are organic compounds that also contain elements such as fluorine, chlorine, phosphorus, nitrogen, sulfur, and metals such as arsenic. This can result in the formation of inorganic compounds such as hydrogen fluoride, hydrogen chloride, phosphorus pentoxide, nitrogen dioxide or other nitrogen oxides, sulfur dioxide, and metal arsenic oxides during the incineration process (Pearson and Magee, 2002). Methods exist for the removal of these materials from the offgases from the process.

One issue with incineration technology is that organic compounds cannot generally be converted completely to carbon dioxide and water. Instead, trace amounts of compounds such as dioxins and furans, as well as other products of incomplete combustion, can be generated during the combustion process and must be controlled by scrubbing and cleaning the offgas using, for example, activated carbon adsorption. Even so, this characteristic of incineration processes has sometimes made it difficult for this technology to gain public acceptance, especially from stakeholders in local communities and environmental interest groups.

Incineration has been used for the destruction of both agents and various types of secondary waste. This report looks primarily at the destruction of agents. The U.S. Army and its contractors have developed very advanced and sophisticated incineration technology for the destruction of the U.S. chemical weapons stockpile. These are high-capacity, fixed facilities designed for destroying munitions in good condition. In general, they are not useful for the destruction of non-stockpile chemical weapons because the deteriorated condition of the weapons will not allow for their disassembly with the installed equipment. Moreover, these facilities are not cost-effective for destroying small quantities of munitions, nor are they readily adaptable for moving from site to site.

The baseline incineration system is shown schematically in Figure 5-4. It includes (1) the reverse assembly of munitions to separate agents, energetics, metal parts, and dunnage; (2) high-temperature combustion of agent in a liquid incinerator; (3) combustion of energetics in a deactivation furnace; and (4) decontamination of metal parts in a metal parts furnace. These three furnaces have secondary combustion chambers to promote complete oxidation of organic compounds. Uncontaminated dunnage is generally disposed of offsite. Each of the three furnaces has elaborate air pollution control systems to remove gaseous pollutants and particles. The unit operations in these systems include caustic scrubbers for acid gas removal, quench towers for gas cooling, venturi scrubbers for particulate removal, fabric filters for elimination of fine particulates and mist, induced draft fans for movement of the exhaust gases, and activated carbon adsorbers for final removal of organic compounds.

Country-by-Country Experience

United States

Before the baseline incineration system had been developed for destruction of the U.S. chemical weapon stockpile, incineration was used to destroy 2,800 metric tons (100 percent agent basis) of chemical weapons, corresponding to 60,000 munitions at the Rocky Mountain Arsenal and the Chemical Agent Munitions Disposal System at the Tooele Army Ammunition Plant, Utah (Sutherland, 1997). The latter installation was used primarily to develop the baseline process, which is now being used to destroy the U.S. stockpile of chemical agents and munitions at four of the eight storage sites in the continental United States. The first completely integrated baseline system was built on Johnston Atoll in the Pacific Ocean. It was successfully used to destroy the stockpiled munitions at that site and has now been closed. The four sites currently in operation are located at Tooele, Utah; Umatilla, Oregon; Anniston, Alabama; and Pine Bluff, Arkansas.

Germany

The German treatment facility at Munster is described by Martens (1997) and by Mihm (2000). It was toured by

[6]See <http://www.sgpproduct.org/Personal%20Use%20Only/GreenCross2.html>.

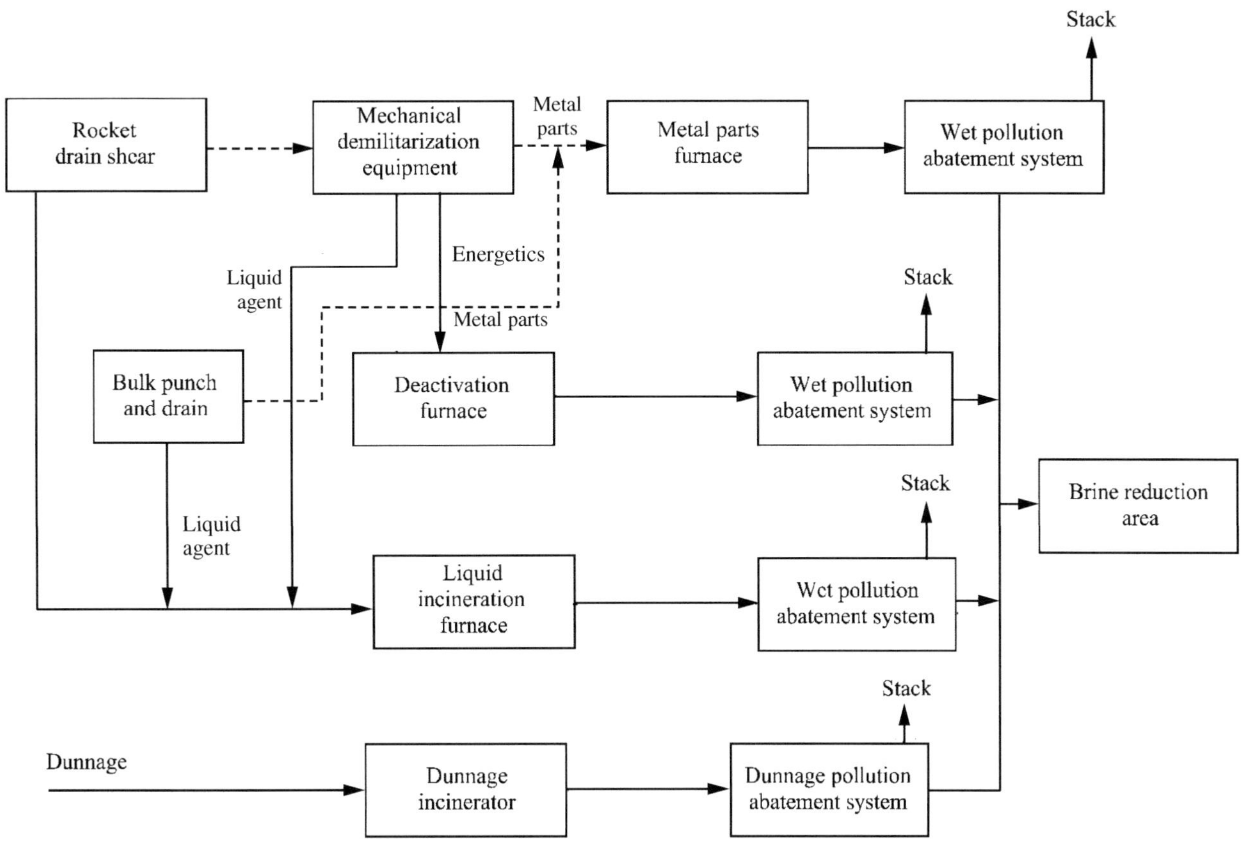

FIGURE 5-4 Block diagram of U.S. baseline incineration system. SOURCE: Pearson and Magee, 2002.

two of the committee members for this report. The facility has a capacity of about 70 metric tons per year, including contaminated waste from operations and metal scrap from munitions. Disassembly of munitions to access the agent involves the use of cutting equipment that requires intensive operator involvement. Solid wastes to be incinerated are placed in polyethylene containers, and liquids are placed in tin boxes. These are fed in a batch fashion to the incinerator. This incinerator has an evaporation chamber, where materials are held for 12 hours at 300°C, and a burn-out chamber, where materials are held for 18 hours at 1000°C. Offgases from these are burned in a main combustion chamber, with a residence time of 2 seconds and a temperature of 1200°C. The flue gases are cleaned extensively before discharging. Solid waste, including arsenic products, is sent to a former salt mine. Aqueous effluent is sent to either a municipal wastewater treatment system or is evaporated.

United Kingdom

The United Kingdom facility at Porton Down employs a licensed hazardous waste incinerator for munitions destruction.[7] This is again a fixed facility, not suitable for moving from site to site.

The weapons are initially stored in open-sided buildings. They are assessed using portable isotopic neutron spectroscopy and x rays to determine the nature of the fill and then placed in commercial freezing isocontainers at –15°C to minimize the polymerization of mustard agent while in storage. Munitions containing phosgene or chlorine are drilled and vented to a caustic scrubbing system to destroy the phosgene or chlorine.

The frozen munitions containing mustard agent are sawed into two parts using either a band saw or a reciprocating saw. This is done in enclosed chambers, with the saws cutting into the cavities containing the mustard agent but above the level of the mustard agent. An operator enters the chamber and pours the agent into a glass container, which is in turn placed in a plastic bucket. The bucket also contains activated carbon to absorb the agent if the glass container leaks. The bucket is

[7]Personal communication between Richard Soilleux, DSTL Porton Down, and Richard Ayen, committee chair, November 14, 2005.

sent to the incinerator for destruction of the mustard. Most of the mustard agent is the HT form.[8]

The incinerator has a continuously fed rotating kiln furnace and a batch-operated, fixed-hearth furnace. Offgases from the furnaces pass to a secondary combustion chamber operating at 1200°C and a 2-second residence time. The furnaces and secondary combustion chambers are generally fired with waste oil. The gases then pass through a liquid quench system that drops the temperature to 200°C. This system is designed to control dioxins and furans. The gases pass into a bag filter that has been precoated with lime or sodium bicarbonate to control acid gases and particulate, then to a caustic scrubber for more acid gas removal. The gases are analyzed continuously at the exit of the scrubber, and the feed to the incinerator is automatically stopped if limits are exceeded. The gases then pass through an induction fan and into a stack.

The bucket containing the mustard agent is placed in the fixed-hearth furnace for destruction. Fuzes and other small or low-energy parts containing energetics are placed in a steel bang box and detonated in the fixed-hearth furnace. Parts containing higher energy energetics are open-detonated. The metal bodies of the munitions are treated thermally in the incinerator and disposed of.

As of January 2006, about 850 chemical weapon rounds were in inventory at Porton Down, and it was expected that these would all be destroyed by April 2007.[9]

Canada

Chemical warfare agents, mainly mustard agent, were transported to DRES after World War II for destruction (Sutherland, 1997). Over 700 tons of mustard agent were destroyed in 1975 and 1976 using alkaline hydrolysis followed by incineration of the hydrolysate and landfilling of the residues.

Another campaign was initiated in 1989 aimed at destroying additional materials, including 12 tons of liquid or thickened mustard agent; 1.5 tons of lewisite; 0.3 tons of nerve agents (G and V classes); 300 tons of contaminated scrap, containers, and ordnance; 4 tons of contaminated waste such as decontaminating solution; and 40 tons of nonhazardous scrap. An incinerator was built for this purpose. Mustard agent was incinerated directly. The mustard-agent-filled munitions were punctured with shaped charges. The agent was drained from the munitions and poured into polyethylene-lined boxes. It was then frozen, transported to the incinerator, and fed through the solid waste feed system.[10] The drained munitions were placed in sealed polyethylene-lined boxes and fed to the incinerator through the solid waste feed system. Nerve agents were neutralized with potassium hydroxide and the hydrolysate was incinerated. The incinerator had a combined liquid plus solid feed rate of 1.5 tons per hour. The incineration campaign took 3 months.

Iraq

An incinerator was built and operated at a desert site in Iraq to destroy mustard agent as part of the 1991-1994 program to destroy the agent production facilities and chemical weapons in Iraq following the Desert Storm military operation (Manley, 1997). All energetics were removed from the weapons. Several techniques were then employed to access the agent in the weapons, including removal of the filling plug when possible, drilling holes, and using shaped explosive charges to create holes. The drained weapons were rinsed, again using various techniques, and then burned. A single incinerator was constructed at the former agent production facility, using an existing tank farm to store agent and prepare it for incineration. Mustard agent was mixed with a particular petroleum fraction in which it was soluble, then with fuel oil. A large brick-lined furnace was constructed in a trench and two burner guns were installed. The combustion gases were diluted with ambient air, passed through a water quench tower, passed through a packed tower irrigated with a 5-10 percent caustic solution, passed through an induction fan, and discharged through a tall stack. The mustard agent feed rate was 300 kg/hr, and the overall campaign destroyed about 400 metric tons of agent.

Russia

A mobile neutralization and incineration system designed to destroy nerve agent was built and operated in Russia beginning in 1980 (Krotovich, 1998). This facility destroyed 300-350 metric tons of chemical agent in a 100-day period.

Evaluation Factors Analysis

Process Maturity

Pearson and Magee (2002) state that more chemical agent has been destroyed by incineration than by any other technology. Table 5-1 shows major chemical weapon destruction

[8]HT is made up of HD, which is bis(2-chloroethyl) sulfide, with 20-40 wt % agent T, bis[2-(2-chloroethylthio) ethyl] ether. HT has a lower freezing point than HD.

[9]Personal communication between Richard Soilleux, DSTL Porton Down, and Richard Ayen, committee chair, January 13, 2005.

[10]At temperatures below 0°C, mustard agent, either distilled or non-distilled, will be a solid and the vapor pressure will be lower than at higher temperatures. This reduces exposure risk from incidents during transport. Transport of mustard- and phosgene-containing munitions in freezer trucks has also been carried out in France (CNN, 2001).

TABLE 5-1 Destruction of Chemical Agents, 1958-1993

Site	Agent	Amount[a] (tonnes)	Method[b]	Dates
Rocky Mountain Arsenal, USA	H	2,786	I	1969-1974
Rocky Mountain Arsenal, USA	GB	3,799	N	1973-1976
Tooele (CAMDS),[c] USA	GB	34.5	I	1981-1986
JACADS,[d] USA – OVT data	GB	36	I	1991
Tooele (CAMDS),[c] USA	VX	7	I	1984
JACADS,[d] USA – OVT data	VX	49	I	1992
JACADS,[d] USA – OVT data	HD	51	I	1992
DRES Canada	H	700	N/I	1974-1976
DRES Canada	H	12	I	1990-1991
DRES Canada	VX, GA, GB	0.3	N/I	1990-1991
DRES Canada	L	1.5	N	
Munster, Germany	H, etc.	70/year	I	1980-present
Shikhany, Russia	GB, GD, H	300	N/I	1980-1990
Shikhany, Russia	VR[e]	30	N/I	1980-1990
Porton Down, U.K.	H	20	I	1970
Nancekuke, U.K.	GB	20	N	1967-1968
Runcorn, U.K.	H	6,000	I	1958-1960
Iraq (UNSCOM supervised)	H	500	I	1992-1993
Iraq (UNSCOM supervised)	GA	30	N	1992-1993
Iraq (UNSCOM supervised)	GB, GB/GF	70	N	1992-1993

[a]U.S. figures based on 2,000-lb tons converted to 1,000-kg tonnes.
[b]I, incineration; N, neutralization.
[c]Chemical Agent Munitions Disposal System experimental facility, Tooele Army Depot, Utah.
[d]Johnston Atoll Chemical Agent Disposal System; data are from preoperational Operational Verification Tests (OVT).
[e]VR is Russian VX.

SOURCE: Adapted from Pearson and Magee, 2002.

projects worldwide from 1958 to 1993. The table lists projects involving incineration, neutralization (including hydrolysis), and a combination of neutralization and incineration.

Most of the projects were performed using either incineration or a combination of neutralization and incineration. Much of the U.S. stockpile has since been destroyed using incineration. This includes all of the 2,030 tons of agent that were in the Johnston Atoll stockpile and much of the 23,437 tons of agent that were at Anniston, Pine Bluff, Tooele, and Umatilla. Incineration of chemical weapons has been or is being used in Canada, Germany, Belgium, Russia, the United Kingdom, and Iraq. In all these countries, incineration was used solely or in part to directly destroy agent. The technology has been used to destroy a variety of agents and energetics in a variety of munitions. Systems of various sizes and degrees of sophistication have been built, ranging from the large, complex systems built for destruction of the U.S. stockpile to the small, considerably less complex system built for the Iraq application.

Process Efficacy/Throughput

Incineration is highly effective in destroying agents and energetics. At Johnston Atoll, the destruction and removal requirement (DRE) for GB and VX was 99.99999 percent and the required DRE for mustard agent was 99.99995 percent.[11] DREs for energetics must be 99.99 percent or greater. The stockpile plants have experienced long periods of downtime, usually following an incident in the facility. Hundreds of workers are employed in the stockpile destruction plants, a large staff in the absolute sense but not unusually large for the tasks being performed; many of the workers are involved in operations peripheral to the incineration-related activities. The process has been used to destroy a wide variety of agents and energetics in a wide variety of munitions. Portable incinerators have been built and are technically feasible, but obtaining permits for operation would be difficult because of public and regulatory acceptance issues.

Process Safety

Dangerous materials are handled in these plants, solids handling operations are extensive, and operations and repairs inherently involve risk. These factors have led to incidents at the U.S. stockpile destruction plants, resulting in downtime. Incidents include releases of agent to the environment, but

[11]The DRE for an incinerator is defined by the EPA as DRE = $[(W_{in} - W_{out})/W_{in}] \times 100\%$, where W_{in} = mass feed rate of a selected organic compound in the waste stream feeding the incinerator, and W_{out} = mass emission rate of the same organic compound present in exhaust emissions prior to release to the atmosphere (40 CFR 264.343).

their causes have been studied extensively and demilitarization facilities personnel have made a serious effort to learn from past incidents (NRC, 2002). The committee anticipates that such incidents will be less frequent in the newer plants (Pine Bluff, Umatilla, and Anniston). Historically, 12 years of operation of two destruction facilities prior to 2002 witnessed three confirmed agent releases to the environment involving the equivalent of a few small drops of agent in total. In comparison, several hundred leaks of stored agent occurred between 1990 and 2000, with the most serious resulting in release to the environment of 78 gallons of mustard agent. Thus, releases from storage of agent have been far more serious than releases from destruction facilities (NRC, 2002). Engineering controls have been put in place to mitigate recognized risks. Interlocks are used extensively. A major safety concern is having unburned agent or products of incomplete combustion exiting the stack. To minimize this risk, agent flow to the burner is stopped if the temperature in the primary combustion chamber drops below 1400°C (Pearson and Magee, 2002).

Note that the risk associated with retrieving the munitions from the bunkers and moving them to the destruction facilities is approximately the same regardless of the destruction process employed, incineration, neutralization, or other.

Public and Regulatory Acceptability in a U.S. Context

Incineration of chemical weapons is applied extensively in the United States and in some foreign countries, such as the United Kingdom, Belgium, and Germany. It is used for both chemical agent and secondary waste, and, overall, it has been an effective technology. In addition, incineration was endorsed as a suitable technology for chemical agent and munitions destruction early on by the NRC (1984), and incinerators were built and operated at four of the Army's stockpile sites. However, for several reasons, incineration has not been well received by public interest groups in the United States, and public pressure resulted in years of delays in obtaining permits, with attendant delays in schedule. This, in turn, caused the investment of millions of dollars in developing alternative technologies for destruction of chemical stockpiles, and such technologies have been selected for use at four of the Army's chemical stockpile sites.

The opposition to the use of incineration is described in detail in a prior NRC report (NRC, 2002). Reasons for it include the perceived instability of the process, the potential for explosion, and the potential for unplanned releases of undesirable compounds. This public opposition to incineration is also evident in the U.S. Environmental Protection Agency's Superfund program for cleanup of hazardous waste sites, where both on-site and off-site incineration were selected less frequently as treatment technologies as the years passed (EPA, 2004). For example, on-site incineration was selected four, seven, six, and four times in 1987, 1988, 1989, and 1990, respectively, for source control for remedial actions but not at all in 1998 through 2002. Similarly, off-site incineration was selected three, nine, nine, fifteen, and twelve times in 1987, 1988, 1989, 1990, and 1991 respectively, and three, two, six, one, and two times in 1998, 1999, 2000, 2001, and 2002, respectively.

The non-stockpile munitions disassembly procedures used in the United Kingdom, Germany, and Belgium require extensive operator involvement, with the attendant risk of exposure to agent. All of these countries consider these technologies to be out of date and are actively considering alternatives.

Secondary Waste Issues

Numerous secondary wastes are produced by incineration processes. Table 5-2 lists the liquid and solid wastes produced by the baseline incineration facility at Tooele (U.S. Army, 1999). This table also shows whether wastes generated are (1) incinerated within the facility, (2) sent to an off-site TSDF, or (3) placed in permitted on-site storage. Treated metal parts from the metal parts furnace (MPF) are described as a high-volume stream that contains the 5X-treated metal bodies of all munitions and agent ton containers.

The U.S. Army does not plan to incinerate the lewisite stored in 10 one-ton containers at the Tooele site. The method of destruction to be employed has not been finalized, but the leading candidate is treatment with sodium permanganate to precipitate a manganese oxide-arsenic solid (Ember, 2006). This solid would be encased in concrete and disposed of in a permitted hazardous waste landfill, with the remaining liquids treated further and injected into a deep well.

Process Costs

Incineration processes, especially as built and operated for the destruction of the U.S. chemical weapons stockpile, are complex and expensive. For example, the Johnston Atoll Chemical Agent Disposal System (JACADS), the first of the baseline plants, cost $254 million to build and $100 million for systemization (Sutherland, 1997). As of 1997, additional costs of over $500 million for operation were expected. It is not known how the costs of an incineration system to be used for non-stockpile chemical weapon destruction would compare with the JACADS costs.

COMPARATIVE EVALUATIONS OF TIER 1 AGENT-ONLY PROCESSING TECHNOLOGIES

Table 5-3 rates the Tier 1 agent-only processing technologies according to the evaluation factors defined in detail in Chapter 3. The ratings scheme is described in more detail in that chapter.

The committee also addressed the subfactors as given in Chapter 3 for each of the five main evaluation factors. The subfactor questions for the Tier 1 international tech-

TABLE 5-2 Tooele Chemical Agent Disposal Facility Waste Streams

Waste Stream	Currently Treated by Incineration?
Waste chemicals	Yes, except for lewisite
Brine reduction area baghouse residue	No, offsite TSDF
Brine salts	No, offsite TSDF
Brine tank sludge	No, offsite TSDF
Neutralization fluid	Yes, except for laboratory operations
Deactivation furnace system (DFS) cyclone residue	No, offsite TSDF
DFS heated discharge conveyor ash	No, offsite TSDF
DFS refractory	No, offsite TSDF
Demilitarization protective ensemble suits	No, permitted long-term storage
Laboratory liquid wastes	No, offsite TSDF
Laboratory solid wastes	No, offsite TSDF
Liquid incinerator refractory	No, offsite TSDF
Liquid incinerator slag	No, offsite TSDF
Miscellaneous metal parts	Yes
Mist eliminator filter/demister candles	No, offsite TSDF
Metal parts furnace (MPF) ash	No, offsite TSDF
MPF metal	Yes
MPF refractory	No, offsite TSDF
MPF residue	No, offsite TSDF
Pollution abatement system (PAS) brine	No, offsite TSDF
PAS brine sump sludge	No, offsite TSDF
PAS quench tower residue	No, offsite TSDF
Plastics	No, permitted onsite storage
Residue handling area baghouse residue	No, offsite TSDF
Spent activated carbon	No, permitted onsite storage
Spent hydraulic fluid	Yes
Spent silver fluoride pads	No, offsite TSDF
Spill cleanup material	Yes
Trash, debris, and protective clothing	No, permitted on-site storage
Waste oil	Yes
Wooden pallets	No, permitted on-site storage

SOURCE: Adapted from U.S. Army, 1999.

TABLE 5-3 Evaluation Factor Rating Comparison of Tier 1 Agent-Only Processing Technologies with U.S. RRS/SCANS

	Evaluation Factors (Rating[a])[b]				
Technology	Process Maturity	Process Efficacy/ Throughput	Process Safety	Public and Regulatory Acceptability in a U.S. Context	Secondary Waste Issues
U.S. Neutralization/hydrolysis (RRS/SCANS)	+	+	+	+	0
Russian two-stage process: neutralization with addition of bitumen	?	+	+	0	–
Incineration	+	+	+	–	–

[a]Legend: +, acceptable; 0, partially acceptable; –, unacceptable; ?, inadequate information.

nologies suitable for non-stockpile agent-only processing are addressed in Appendix C in Tables C-1 through C-5. These tables provide a convenient side-by-side means for comparing some specific aspects of these international technologies with the NSCMP equipment based on the available data and the expert judgment of the committee.

FINDINGS AND RECOMMENDATIONS

Finding 5-1. In the Russian context, the destruction efficiency for the Russian two-stage bitumen process is sufficiently high, the reactions involved are irreversible, and the toxicity of the bitumen-salt product from the Russian two-stage process is sufficiently low.

Finding 5-2. The Russian two-stage neutralization/bitumenization process generates a large amount of secondary wastes that in the United States would require being landfilled and monitored indefinitely. The lack of information on the long-term stability of these wastes and the leaching of toxic materials from them (see "Secondary Waste Issues") is a cause for concern.

Recommendation 5-1. The U.S. Army should not expend any resources on further evaluation of the Russian two-stage neutralization/bitumenization process for application in the United States.

Finding 5-3. Incineration is a highly developed and well-proven technology for the destruction of chemical agent and various types of secondary waste. Public acceptability of the technology is poor within the United States but adequate in most other countries.

Finding 5-4. The reverse assembly process used in the U.S. baseline incineration system is not applicable to non-stockpile chemical weapons because of the deteriorated conditions of the munitions.

Finding 5-5. Incineration of non-stockpile materials is used in the United Kingdom, Belgium, and Germany in conjunction with human handling and remotely controlled cutting of munitions. Such operations are labor intensive and could cause operators to be exposed to agent.

Finding 5-6. Although incineration is a robust, safe, and proven technology for the destruction of chemical agent, it is not the best technology for non-stockpile materials because access and acceptability are issues.

Recommendation 5-2. The U.S. Army should consider the difficulty of obtaining public acceptance when comparing incineration with other non-stockpile agent destruction technologies for use in the United States.

REFERENCES

Neutralization and Hydrolysis

GSN (NTI: Global Security Newswire). 2005. Russian CW Destruction Facility Nears Completion, October 27. Available online at <http://www.nti.org/d_newswire/issues/2005_10_27.html>. Last accessed March 6, 2006.

Kawataki, H., H. Shimoyama, Y. Yamane, and C. Furukawa. 2005. Destruction of OCW in Glass Beer Bottles at Samukawa. Available online at <http://www.dstl.gov.uk/news_events/conferences/cwd/2005/proceedings50.pdf>. Last accessed March 6, 2006.

Litovkin, V. 2005. Russia takes another step towards scrapping its chemical weapons. Available online at <http://en.rian.ru/analysis/20051201/42278210.htm>. Last accessed March 6, 2006.

Manley, R.G. 1997. "UNSCOM's experience with chemical warfare agents and munitions in Iraq." Pp. 241-254 in The Challenge of Old Chemical Munitions and Toxic Armament Wastes. T. Stock and K. Lohs, eds. Oxford, England: Oxford University Press.

Pearson, G.S., and R.S. Magee. 2002. "Critical evaluation of proven chemical weapon destruction technologies (IUPAC Technical Report)." Pure and Applied Chemistry 74(2): 187-316.

Sutherland, R.G. 1997. "The destruction of old and obsolete chemical weapons: Past experience." Pp. 141-155 in The Challenge of Old Chemical Munitions and Toxic Armament Wastes. T. Stock and K. Lohs, eds. Oxford, England: Oxford University Press.

Russian Two-Stage Process

Bechtel (Bechtel International, Inc.). 1996. Joint Evaluation of the Russian Two-Stage Chemical Agent Destruction Process, Final Technical Report: Phases 1 & 2, Revised, July. Alexandria, Va.: Defense Special Weapons Agency.

Ember, L. 2003. "Ridding Russia of chemical weapons." Chemical & Engineering News 81(48): 28-31.

Ember, L. 2005. "The Shchuch'ye dilemma." Chemical & Engineering News 83(45): 19-23.

Khripunov, I., and G.W. Parshall. 1999. "U.S. assistance to Russian chemical weapons destruction: Identifying the next steps." Available online at <http://cns.miis.edu/pubs/npr/vol06/64/krip64.pdf>. Last accessed March 6, 2006.

Krotovich, I.N. 1998. "Development of CW destruction technologies in Russia." Pp. 89-93 in Proceedings of the International CW Destruction Symposium, Munster, Germany, March 22-25, 1998. V. Starrock and A. Krippendorf, eds. Munster, Germany: Munster Expo 2000 Committee.

Pearson, G.S., and R.S. Magee. 2002. "Critical evaluation of proven chemical weapon destruction technologies (IUPAC Technical Report)." Pure and Applied Chemistry 74(2): 187-316.

Petrov, V.G., and A.V. Trubachev. 2000. "On certain problems of technical and ecological safety during chemical weapons destruction in the Udmurt Republic." Proceedings of Munster 2000, The 2nd International Symposium, Destruction of Chemical Weapons—Technologies and Practical Aspects, July 30-August 3, 2000. Munster, Germany: Munster Expo 2000 Committee.

Radilov, A.S. 2004. "Toxicological and hygienic aspects of the problem of chemical detoxification of organophosphorus chemical warfare agents." Available online at <http://www.dstl.gov.uk/news_events/conferences/cwd/2004/proceedings25.pdf>. Last accessed March 6, 2006.

Ryan, S., and P. McNelly. 2003. "Eliminating Russian nerve agent weapons." Proceedings of the 2003 Chemical Weapons Demil Conference. Porton Down, England: Defence Science and Technology Laboratory.

Incineration

CNN (Cable News Network). 2001. WWI munitions transferred safely. Available online at <http://archives.cnn.com/2001/WORLD/europe/04/16/Vimy.munitions.02/index.html>. Last accessed March 6, 2006.

Ember, L.R. 2006. "Chemical weapons deadline at risk." Chemical and Engineering News 84(16): 27-30.

EPA (U.S. Environmental Protection Agency). 2004. Treatment Technologies for Site Cleanup: Annual Status Report (11th Edition), EPA-542-R-03-009, February. Available online at <http://www.clu-in.org/download/remed/asr/11/asr.pdf>. Last accessed February 14, 2006.

Krotovich, I.N. 1998. "Development of CW destruction technologies in Russia." Pp. 89-93 in Proceedings of the International CW Destruction Symposium, March 22-25. V. Starrock and A. Krippendorf, eds. Munster, Germany: Munster Expo 2000 Committee.

Manley, R.G. 1997. "UNSCOM's experience with chemical warfare agents and munitions in Iraq." Pp. 241-254 in The Challenge of Old Chemical Munitions and Toxic Armament Wastes. T. Stock and K. Lohs, eds. Oxford, England: Oxford University Press.

Martens, H. 1997. "The German programme for the disposal of old chemical weapons." Pp. 166-178 in The Challenge of Old Chemical Munitions and Toxic Armament Wastes. T. Stock and K. Lohs, eds. Oxford, England: Oxford University Press.

Mihm, J. 2000. "Activities in the field of chemical weapons agent disposal in Germany." Proceedings of Munster 2000, The 2nd International Symposium, Destruction of Chemical Weapons—Technologies and Practical Aspects, July 30-August 3. Munster, Germany: Munster Expo 2000 Committee.

NRC (National Research Council). 1984. Disposal of Chemical Munitions and Agents. Washington, D.C.: National Academy Press.

NRC. 2002. Evaluation of Chemical Events at Army Chemical Agent Disposal Facilities. Washington, D.C.: The National Academies Press.

Pearson, G.S., and R.S. Magee. 2002. "Critical evaluation of proven chemical weapon destruction technologies (IUPAC Technical Report)." Pure and Applied Chemistry 74(2): 187-316.

Sutherland, R.G. 1997. "The destruction of old and obsolete chemical weapons: Past experience." Pp. 141-155 in The Challenge of Old Chemical Munitions and Toxic Armament Wastes. T. Stock and K. Lohs, eds. Oxford, England: Oxford University Press.

U.S. Army. 1999. Incineration as a Land Disposal Restriction Treatment Technology for Chemical Agent Associated Waste. Available online at <http://www.hazardouswaste.utah.gov/CDS/OTHER/UCAR/UCAR_VOLUME_2/Background%20Document%20H.pdf>. Last accessed January 17, 2006.

Zanders, J.P. 1997. "The destruction of old chemical weapons in Belgium." Pp. 197-230 in The Challenge of Old Chemical Munitions and Toxic Armament Wastes. T. Stock and K. Lohs, eds. Oxford, England: Oxford University Press.

6

Tier 2 International Technologies for Munitions and Agent-Only Processing

INTRODUCTION

Based on its information gathering, the committee assigned Tier 2 status to those technologies that did not at this time warrant the more extensive evaluation given to Tier 1 technologies in Chapters 4 and 5. As described in Chapter 1, the Tier 2 technologies are of two basic types. The first comprises those technologies that may eventually prove to be applicable to Non-Stockpile Chemical Materiel Program (NSCMP) projects but that are still at an early stage of development for such applications. The second type includes those technologies that were tried out in operations for destroying recovered CWM but have so far not proven to be satisfactory for various reasons. Both types are presented below in alphabetical order, first the munitions processing technologies and then the agent-only processing technologies. These technologies (for the reasons given above) do not warrant the investment of U.S. Army resources to develop a treatment process for the non-stockpile program, whether the vendor of the technology is a U.S. entity or a foreign entity. The remainder of this chapter describes these technologies and provides reasons for this conclusion.

TECHNOLOGIES FOR MUNITIONS PROCESSING

Acid Digestion Process

Initial development of the acid digestion process (ADP) was carried out by the Société Nationale des Poudres et Explosifs (SNPE) Propulsion in France (Gaudre et al., 2001).[1] The development effort was continued by Battelle, and the Shaw Group is now apparently involved (Soilleux, 2005). The discussion of the ADP in this section is in part based on published documents and on information obtained during visits to overseas sites. The committee asked for information from the vendor that is currently developing this technology, but because any such information would have been classified as "export confidential" under U.S. regulations, it was not possible for the committee to accept this information. It is known that the NSCMP is evaluating this technology in detail and it is possible that operational or demonstration trials will be performed. The assessment contained in this report was based on the information in the public record and, to a significant extent, on the opinions of experts in chemical demilitarization operations overseas who have additional knowledge of the technology. The committee recognizes that information that could have been provided by the vendor might have altered the conclusions that were reached. However, the report *Possible Replacement Technologies for Operation Pongee After 2007* stated as follows: "More recent communications with the French prime contractor for Secoia reveal that they have revisited all candidate technologies and again rejected acid digestion on the grounds of safety and an incomplete process" (Soilleux, 2005, p. 14). Based on this information and on other sources, the committee concludes it is unlikely that acid digestion will prove to be a technology that is superior to other currently available technologies.

ADP uses 7M nitric acid to access munitions contents, destroy munition bodies, and oxidize the fills. The acid digests the steel that contains the hazardous substances, renders the fuzing systems inoperable, and decontaminates the chemical agents. The acid neutralizes/reacts with the agent by entering the agent cavity through the fuze. A significant amount of heat is generated by the reaction of the acid with the steel and must be removed. The system comprises the following parts:

- Reactor vessel where the munition is placed,
- Acid storage tank with recirculation pump,
- Acid heating and cooling systems,

[1] Unless otherwise noted, information for this section was obtained from Edward Doyle, Alternative Systems Demonstration and Evaluation Group Leader, PMNSCMP, "Acid digestion of GTRs: Results and Path Forward," briefing to the committee, September 7, 2005.

- Air emissions control system,
- Gas hold-up tank, and
- Caustic and permanganate scrubbers to treat the gaseous effluents.

The munition is placed in a sealed reaction vessel and nitric acid is circulated through the reactor, with the exothermic heat of reaction removed by a heat exchanger in the recirculation loop. The reaction is continued until both the agent and energetic fills are neutralized. The reactor is then flushed and drained and the munition removed. The unit is mounted on skids, and the reactor chamber can be tailored to accommodate a wide variety of shapes and sizes of material.

The result of the treatment is a liquor that requires further treatment for recovery or disposal. The liquor includes spent nitric acid, permanganate and sodium hydroxide wastes, and cleanup rinse waters. It also contains the explosive in the munition, which is, theoretically, untreated and must be removed by filtering.

Recovered munitions containing phosgene, mustard agent, or chloropicrin are reported to have been destroyed using ADP. Lab tests are reported to have shown it can also neutralize diphosgene, hydrogen cyanide, and nerve agents. The reported energetic material and chemical warfare fills that the process can be used to treat are given in Table 6-1.

The fate of the energetics has not been firmly established. Although it is claimed that the energetics can be filtered out of the waste liquid, there is a potential issue with energetics contaminating the downstream liquid system.

Advantages of ADP over the explosive destruction system (EDS) may be greater throughput capacity, smaller volume of waste produced, and no requirement for explosive containment. Disadvantages appear to be lower destruction efficiencies (based on tests conducted by Battelle), a requirement for offgas treatment, and safety concerns.

TABLE 6-1 Energetic Materials and Chemical Warfare Fills Treatable by the Acid Digestion Process

Energetic Materials	Chemical Warfare Fills
Picric acid	Phosgene
TNT	Diphosgene mustard agent
RDX	Chloropicrin
Dinitrobenzene	Hydrogen cyanide
Hexanitrodiphenylamine	Tin tetrachloride (smoke)
Dinitrotoluene	Arsenic trichloride (smoke)
Black powder	Nerve agents VX and G series
Lead styphanate	Arsine
Mercury fulminate	

SOURCE: Edward Doyle, Alternative Systems Demonstration and Evaluation Group Leader, PMNSCMP, "Acid digestion of GTRs: Results and path forward," briefing to the committee, September 7, 2005.

The report *Possible Replacement Technologies for Operation Pongee After 2007* (Soilleux, 2005, p. 13) states as follows:

> France has tested the acid digestion equipment as part of their Secoia project. From the very little information available it appears that, although the method worked as planned, there was an explosive incident with a fuze/small burster which was sufficiently serious to lead to the abandonment of the technology for Secoia. The issue of sensitization and/or functioning of live fuzes needs very careful consideration before trials are carried out at Porton. Fired fuzes are the most problematic except for the time and combustion fuzes used in 25pdrs [pounders] which cannot function when wet.

The information surrounding the abandonment of ADP by the French and an explosive incident associated with it was confirmed by representatives of some European organizations involved with chemical demilitarization who supplied information to this committee.[2]

Assessment of Potential Advantages

ADP appears to be theoretically suitable for processing munitions that are damaged or corroded. A munition of any size can be processed, in theory, assuming the reactor is large enough to contain it. According to published claims, no explosion is possible, the process may be considered safe, and the reaction vessel does not have to be designed to withstand an explosion. However, in view of undocumented reports of an incident in France, this claim should be examined carefully.

Assessment of Potential Disadvantages

Several disadvantages concerning ADP can be envisaged, but it may be possible to apply safeguards. However, given the lack of detailed information, there are some potential issues:

- The use of highly concentrated nitric acid poses a significant hazard for personnel that must not be underestimated or minimized.
- The munition must be handled, which presents another hazard for personnel.
- It is not clear on what basis the reaction is considered complete. It may be that a reaction time is specified, or it may be that some objective criterion is used, such as an analysis.
- Side reactions could generate an unstable compound during the reaction with nitric acid. Such unstable

[2]Meetings with Ralf Trapp, Jeff Osborne, and Jerzy Mazur, Organisation for the Prohibition of Chemical Weapons, The Hague, The Netherlands, January 18, 2006, and Herbert DeBischopp and Michel Lefebvre, Belgian Royal Military Academy, Brussels, Belgium, January 19, 2006.

reaction by-products have caused numerous explosions elsewhere in the chemical industry. It may be difficult to eliminate this concern. Reports from a variety of sources on a least one (and possibly more) explosion incidents with the ADP were noted during information-gathering visits to European organizations involved with chemical demilitarization (Soilleux, 2005).[3]

- It is not clear whether this technology is suitable for all agents that require treatment or only for some.
- It is not clear where the waste from the reactions with the agent and explosive would go, the extent to which this waste is hazardous, and what secondary treatment might be required. This waste stream, containing concentrated nitric acid and other wastes, would be difficult to handle.
- ADP requires an offgas treatment system of some complexity.
- The explosives filtered from the liquid waste would have to be destroyed in some manner, presumably by incineration on-site.

Other issues may exist that can only be discovered by a detailed review of the process, which was not possible given its current "export confidential" classification. In short, notwithstanding its apparent simplicity, the issues pertaining to ADP include the use of concentrated nitric acid, the potential for the generation of unstable nitrate compounds, the probable difficulty with handling the secondary waste streams, and the probable requirement to filter out explosives from the waste liquid and then treat the explosives. The French purportedly abandoned ADP after at least one explosive event and after having invested considerable effort to develop it. All of these considerations make it likely that this technology should not be pursued. However, as stated earlier, if more detailed information could be obtained from the vendor without having to treat it as confidential, the committee might be able to change its recommendation.

Bulk Vitrification Process (GeoMelt)

Vitrification is the process of melting materials to produce a glass or glass-like substance. The GeoMelt Division of AMEC Earth and Environmental, Inc., offers a bulk vitrification process for waste treatment that it calls In-Container Vitrification (ICV). AMEC is an English company, and the technology received a favorable evaluation in Soilleux (2005) for possible use in the United Kingdom for the destruction of recovered CWM.

[3]Meetings with Ralf Trapp, Jeff Osborne, and Jerzy Mazur, Organisation for the Prohibition of Chemical Weapons, The Hague, The Netherlands, January 18, 2006; and Herbert DeBischopp and Michel Lefebvre, Belgian Royal Military Academy, Brussels, Belgium, January 19, 2006.

ICV is being considered for the treatment of low-radioactivity tank waste now stored at the Department of Energy's (DOE's) Hanford, Washington, site (CH2MHILL, 2005). At Hanford, the bulk vitrification technology might be used to supplement the operation of the main tank waste treatment plant, possibly treating up to 42 percent of the 53 million gallons of liquid and solid wastes now stored in 177 large underground tanks. The ICV technology will use vitrification technology for the treatment of less active waste. The main waste treatment plant also uses vitrification technology and can treat more active waste.

The Washington State Department of Ecology approved a RCRA research, development, and demonstration permit for the ICV pilot plant (CH2MHILL, 2005; WNO, 2005).[4] The permit is very specific, allowing DOE to do testing for no more than 400 operating days and to treat no more than 300,000 gallons of waste from Hanford's tank No. S-109. The building must be torn down when testing is completed. The testing will be done at full scale, i.e., using the same size container as would be used in the production facility.

Laboratory testing at various scales, up to an engineering scale that was 1/16 the size of the planned full-scale unit, has been under way since about 2003. The objective of this testing has been to identify additives that will promote the formation of a suitable final glass waste form. It is important that the waste formed exhibit minimal leaching of radioactive components and other contaminants of concern, such as RCRA heavy metals (Buelt et al., 1987; Loehr et al., 1992; Thomas and Treat, 2001).

The waste to be treated is mixed with glass-forming additives and local silica-rich earth, then dried, using a batch-mode rotary mixer/dryer and indirect steam heating (WDOE, 2005). The mixer/dryer has a capacity of 10,000 liters (2,645 gallons) at a fill fraction of 45-50 percent. During drying, moisture is removed from the offgases, and the dried gas is added to the main offgas treatment system for additional emission control.

The waste container is expected to be a steel box approximately 3 m (10 ft) high, 2.4 m (8 ft) wide, and 7.3 m (24 ft) long. Before waste is placed in it, the container will be lined with insulating board, sand, and a layer of castable refractory. The refractory will be in contact with the waste. A layer of melt-initiating graphite and soil will be placed over the refractory in the bottom of the container. The container will have one or more ports for sampling the vitrified waste after it has cooled.

A steel lid with attached electrodes will then be placed on the container. The lid is attached to the container using bolted flanges and a refractory gasket. The lid has several ports for waste addition, electrode connections, venting, sampling, and addition of postvitrification material. Some or all of the waste is placed in the container (see next paragraph), and electric

[4]See NRC (2002) for background information on RCRA research, development, and demonstration permits.

power is applied. The waste is heated to about 1300°C over about 140 hours, forming a molten material. After passing through a high-efficiency particulate air (HEPA) filter, ambient air flows into the top of the container to maintain flow to the offgas treatment system, cool the vitrification offgases, and provide thermal protection to HEPA filters in the offgas treatment system. Vitrification offgases are vented under induced draft and flow to an offgas treatment system. As melting occurs, the depth of the waste decreases.

After all the waste has been melted, the melt is allowed to cool, forming the vitrified glassy material. Soil is added to the container so that it is at least 90 percent full. After the vitrified waste is sampled and tested, all connections to the container are removed. The container, still containing the electrodes, is taken away for storage or burial. During processing the volume of waste shrinks by 50 percent or more. Used personal protective equipment (PPE) and other secondary wastes can typically be recycled to the next batch of waste being treated, minimizing the waste.

Offgas treatment is extensive and involves use of sintered metal filters, quench systems, venturi scrubbers, a condenser, a mist eliminator, an offgas heater, parallel HEPA filters, a carbon filter for radioactive iodine removal, a baghouse, and a selective catalytic reduction unit; a packed tower scrubber system is used as a backup.

GeoMelt claims that its base technology has been granted a national Toxic Substances Control Act permit for the treatment of wastes containing up to 1.7 percent polychlorinated biphenyls (PCBs) (Campbell et al., 2005). Bulk vitrification was demonstrated for the treatment of waste from the Rocky Flats Environmental Technology site near Denver, Colorado.[5] About 21,500 pounds of waste containing PCBs and low-level radioactivity were shipped from Rocky Flats to Waste Control Specialists near Andrews, Texas, and treated.

Two melts were conducted, with each melt taking 3 days and producing monolithic blocks about 80 cubic feet in size, a reduction in waste volume of more than 50 percent. The refractory and insulation from the first batch were reused for the second batch. PCBs, trichloroethylene, and perchloroethylene were reduced to below detection limits (not reported).

Information on gaseous emissions was likewise not reported. However, it is possible and perhaps likely that some of these organic compounds were thermally desorbed from the solid waste and that the extent of their destruction depended on the effectiveness of the offgas treatment system.

The technology was also demonstrated at Waste Control Specialists for the treatment of 9,575 pounds of another waste, source not indicated, that also contained PCBs and low levels of radioactivity.[6] As received, the waste was about 25 percent oil and aqueous fluid and 75 percent sludge. Zeolite, bentonite, and soil were added to the waste, and treatment was carried out in five batches. The average batch treatment time was 53 hours. PCBs were reduced to below detection limits (not reported).

To the committee's knowledge, the process has never been used on chemical weapons or agent. If it were to be used for this purpose, the high temperatures and long residence times would be expected to result in complete removal of any agent from the solid phase. However, some agent might be thermally desorbed from the feed material, which would mean depending on the offgas treatment system to destroy the balance of the agent. This could cause problems for public and regulatory acceptability in the United States. Also, large amounts of electricity are consumed, which might hinder public acceptance.

GeoMelt's in situ technology has been tested on a small scale for the destruction of energetics, such as HMX, RDX, and TNT, in soil. These tests resulted in a vitrified product with no detectable explosives (Osborne, 2003; Campbell et al., 2005).

About 65-90 percent of heavy metals such as lead, cadmium, and arsenic are retained in the melt. The fugitive arsenic is captured in the offgas treatment system, "treated by standard arsenic treatment technology" (unspecified), and returned to the next melt (Osborne, 2003, p. 3).

While technically interesting, bulk vitrification has to date not been tested on either munition or agent destruction. It was therefore judged by the committee to not warrant listing as a Tier 1 technology for this report.

Firing Pool

The firing pool is a large pool filled with an aqueous solution that can neutralize the agent in a chemical weapon (Guir, 1997). Guir (1997) says that "the firing pool technique discussed here was designed by Société Nationale des Poudres et Explosifs (SNPE) Ingénierie and based on an unpublished laboratory study at the Centre d'Etudes du Bouchet (CEB)" (Guir, 1997, p. 161).

Explosive charges are placed on the munition, which is then submerged into the center of the pool, and the explosive charges are detonated. The munition is converted into fragments and the burster is exploded. The pool is designed to withstand multiple explosions, and it is expected that one weapon can be destroyed every 15 minutes. From laboratory-scale experiments, it was concluded that a firing pool with a 12-meter diameter and a 6-meter depth, filled with approximately 500 cubic meters of aqueous decontamination

[5]Brett Campbell, GeoMelt Division, AMEC Earth and Environmental, Inc., "Vitrification of Rocky Flats depleted uranium mixed waste (fact sheet)," project summary provided to the committee, December 22, 2005.

[6]Brett Campbell, GeoMelt Division, AMEC Earth and Environmental, Inc., "Mixed TSCA low-level radioactive waste treatability demonstration (fact sheet)," project summary provided to the committee, December 22, 2005.

solution, could be used to destroy about 100 metric tons of mustard agent (Guir, 1997).

A concern with the process is that mustard agent that has been contained in a munition for 70 or more years is likely to have polymerized, becoming viscous and sticky. This sticky mustard agent is likely to adhere to the fragments of the munition. The fragments are expected to be large and few in number. As a consequence, it is expected that agitation of the pool would not be sufficient to cleanse the agent from the fragments, necessitating the use of a downstream fragment cleaning system. It is also expected that large amounts of agent would escape from the surface of the pool, requiring the use of some means of destroying the agent in the vapor phase, with an attendant air pollution control system.

No mention of the development of the firing pool technology past the bench scale was found in the literature. The firing pool technology was not considered further by the committee because it had not been further developed and because of the problems that were uncovered during the bench-scale studies.

TECHNOLOGIES FOR AGENT-ONLY PROCESSING

Biological Approaches

Introduction

Biological approaches for the destruction and detection of chemical agents are being developed for applications such as the mitigation or remediation of contaminated media, sensors to detect the presence of agent, degradation of agents using enzyme extracts, as well as a few other less mature applications. None of these approaches has been fully developed into a technology on a large scale. However, several hold promise to complement the large-scale physical/chemical techniques used for the destruction of non-stockpile chemical weapons. For the nerve agents and blister agents (vesicants), microorganisms and some of the mechanisms for degrading/neutralizing the agents have been identified. Several enzymes from bacteria and fungi have been identified that are capable of degrading a wide range of organophosphate compounds. A considerable effort has been expended on the organophosphate-degrading enzymes owing in part to their potential utility for pesticide destruction. None of these concepts was developed sufficiently to justify its inclusion in this report as a Tier 1 technology.

Biodegradation

Naturally occurring organisms have been found that can degrade a range of the chemical agents. Degradation in soils has been documented. Bioreactors have been developed for select chemical agents, either parent compounds or various degradation products of parent compounds. None of these have been expanded beyond the pilot scale, and most exist only at the bench scale. The stability of the microbial cultures remains a challenge. Much work has been done on the isolation of key enzyme systems. Although some studies exist that demonstrate the performance of these enzymes in aqueous suspensions, a significant effort has been expended on developing immobilized systems. These have been in the form of cryoimmobilized beads, packed columns, and impregnated cotton materials. Bioremediation technologies for soils contaminated by chemical agents were discussed at the Third International Workshop "Biotechnological Approaches to Chemical Weapon Destruction," in Saratov, Russia, in 2000. The feasibility of using naturally occurring microorganisms of genetically modified bacteria has been proven, but no field-scale operations have been reported. There is also the potential to use a bioremediation/phytoremediation approach for treating contaminated soils. Patents have been filed for two degradation processes: one uses enzymatic processes and the other a reactor system for phosphonate degradation.

Biomaterials

Some materials have been developed for the protection of personnel and farmers who might be exposed to chemical agents or pesticides. The clothing that is currently available has an absorptive polyurethane layer impregnated with activated carbon, which offers protection but does not neutralize/degrade the agents. It is expected that organophosphate degrading enzymes will be incorporated into this material. Enzyme-containing materials are also being developed for self-decontaminating clothing and surfaces. Because the preparation and purification of enzymes is expensive, work is focusing on immobilization techniques that can maximize enzyme performance and longevity. A brief summary follows:

- *Pseudomonas diminuta* (organophosphate hydrolase) immobilized on nylon performed for weeks without leaching (Caldwell and Raushel, 1991a, b). The type of support material was a key factor in the successful degradation or organophosphates (Havens and Rase, 1991).
- Taking advantage of cloning methods, an *E. coli* with surface-expressed organophosphate hydrolase was immobilized on cotton used in fabrics and filters (Ritchens et al., 2000; Wang et al., 2002). Grimsley et al. (2001) demonstrated that the tremendous capacity of cotton to absorb the nerve agent-degrading enzymes made it appropriate for use as decontaminating towelettes, gauze, swabs, bandages, and wound dressings.
- A variety of bioplastics and enzyme-polymer composites for use as reactive monoliths, foams, fibers, wipes, and coatings have been developed (Kline et al., 2000; Gill and Ballesteros, 2000a, b). In this form, the enzymes were found to maintain stability under

normally denaturing conditions. Easily prepared enzyme-containing polyurethanes have a wide range of properties (Braatz, 1994). Foams and gels can be prepared, but there has been variable success with these due to long-term stability problems. However, a sol-gel product has been shown to retain high activity and to have good stability (Gill and Ballesteros, 1998, 2000c).

Sandia Laboratories, in collaboration with EnviroFoam, has commercialized a product that is effective in the decontamination of materials exposed to chemical agents as well as to some pathogenic viruses and bacteria.[7] This product consists of several chemicals, including some enzymes, and has been tested at several sites for different types of contamination.

Biosensors

Several biosensors are currently being developed primarily for detecting the release of agents into the air. However, some could be adapted for detecting agent in water or soil media. A project funded by DOE's Initiatives for Proliferation Prevention program has resulted in an analyzer capable of detecting organophosphates, carbamates, and other inhibitors of butyrylcholinesterase. The electrochemical biosensor analyzer for detection and discrimination of different neurotoxins can analyze water, soil, and food samples, and has potential for air control and medical applications.

Defense Science and Technology Laboratory Electric Cylinder Furnace

Exposure to high temperatures was used to destroy mustard agent and mustard agent/lewisite mixtures in a bench-scale study at the DSTL, Porton Down, England (Anderson et al., 2003). The agent or agent mixture was evaporated and mixed with an artificial air containing 79 percent nitrogen by volume and 21 percent oxygen.

Whether or not the amount of added air was less than or more than the stoichiometric amount for full oxidation of the agent(s) was not stated. The agent/air mixture was passed through quartz tubes contained within three close-coupled electric furnaces. The first oven was maintained at 600°C or 750°C, the second at 1100°C or 1200°C, and the third at 800°C. Product samples were collected and analyzed. All agent was reportedly destroyed. $AsCl_3$, 1,2,4-trithiolane, 1,3,5-trithiane, cyclic octa-atomic sulfur, and $As(SC_4H_9)_3$ were identified in the products. Trace quantities of dioxins were also detected. The investigators concluded the results were sufficiently encouraging to warrant "consideration of the method as a means of dealing with the problem posed by the old and abandoned chemical munitions. The process appears to be safe, reliable and robust" (Anderson et al., 2003, p. 15).

Without knowing the agent/air mixture ratio, it is not possible to determine whether the destruction process was similar to combustion or to pyrolysis. Only one paper was found, and no further work was carried out by DSTL. This technology must be considered to be at a very early stage of development and therefore presently of little interest for non-stockpile application.

Electrochemical Oxidation

An earlier study (NRC, 2001a) for the U.S. Army NSCMP concerned with the treatment of liquid neutralent wastes produced by NSCMP disposal operations examined two electrochemical processes, the silver Ag(II) process, which was developed by the firm AEA Technologies in the United Kingdom, and the cerium Ce(IV) (or "CerOx") process. Although the latter process was developed in the United States, the two processes are similar, and both are included here for completeness. A description of these two processes from the earlier NRC report follows:

Ag(II) Process

> This process has been patented for oxidizing organic wastes using Ag(II), an unstable form of silver and one of the strongest oxidizing agents known. Any carbon in the waste stream is completely oxidized to carbon dioxide with traces of carbon monoxide. Other elements end up as salts (e.g., fluorines to fluorides, sulfur to sulfates). Chlorine precipitates out with the silver as silver chloride. The process is operated at 90°C and at atmospheric pressure.
>
> A solution of silver nitrate in 8-molar nitric acid is electrolyzed to produce the Ag(II) cations at the anode of a commercially available electrochemical cell. A semi-permeable membrane separates the anode and the cathode compartments of the cell to prevent mixing of the anolyte and catholyte solutions but allowing the passage of cations and water across the membrane.
>
> The anolyte and catholyte solutions form two separate recirculating loops. The anolyte solution is circulated through the reaction vessel into which the organic wastes are introduced. Solids formed in the anolyte loop are removed by a hydrocyclone. In the cathode loop, the nitric acid is reduced to nitrous acid and water. This solution is passed through a nitrogen oxide reformer to regenerate nitric acid. Off-gases are passed through a scrubber. If no chlorine is present, the silver ions are recovered and recycled to the anolyte loop. (NRC, 2001a, p. 26)

CerOx Process

> The CerOx process is similar to the Ag(II) process except that it uses 0.8M Ce(IV) solution in 3-molar nitric acid at 100°C to oxidize and destroy organic compounds. Unlike Ag(II), Ce(IV) is stable. The Ce(IV) is produced and regener-

[7]See <http://www.envirofoam.com/EasyDecon>.

> ated by the electrolysis of Ce(III) in a bipolar electrochemical cell, which the vendor calls a "T-cell."
>
> The system has two circulating loops, one for the anolyte solution and one for the catholyte solution. In the anolyte loops, Ce(III) is oxidized to Ce (IV) in the T-cell and passed through the reaction chamber where the organic wastes are introduced gradually. Carbon is converted to carbon dioxide; chlorine compounds are converted to elemental chlorine, which is scrubbed and converted to hypochlorite; sulfur and other elements are converted to salts, such as sulfates. These salts remain in anolyte solution, which must be periodically replaced as the concentration of the salts increases.
>
> The catholyte loop provides the second electrode for the electrolysis. The nitric acid in this loop is reduced to nitrous acid and then reformed back to nitric acid and nitric oxide. Water is produced in the process, but much of it is removed by evaporation because the operating temperature is very close to the boiling point (100°C).
>
> The CerOx process uses very few reactants, principally nitrate (which is recycled), nitric acid, and sodium hydroxide scrubbers to treat off-gases. The biggest cost is for electrical power to operate the electrolysis T-cells. (NRC, 2001a, p. 28)

The Ag(II) process was also examined as a potential technology for the disposal of assembled chemical weapons at two of the U.S. stockpile storage sites. A report on the Assembled Chemical Weapons Assessment program demonstration phase testing conducted for this process (NRC, 2001b) listed eight findings and four recommendations, which pointed to a number of observed shortcomings with regard to the application of this technology to chemical demilitarization at that time: continuing major design changes during the testing period, migration of organic material across the electrochemical cell membranes, and generation of new energetic compounds in the course of processing.

A further disadvantage of the Ag(II) process is that large quantities of silver (a toxic heavy metal) and nitric acid (a corrosive) are required, along with the attendant potential for production of toxic emissions and effluents.

The CerOx process avoided some of the difficulties of the Ag(II) process in that cerium is much cheaper and less toxic than silver. However, the most serious disadvantage for the CerOx process, at least at the time of the report (NRC, 2001a), was that it was found to not be as mature a technology as the Ag(II) process, and it had never been tested with any neutralents. As with the Ag(II) process, it also uses large amounts of nitric acid and thus rated poorly in terms of pollution prevention criteria.

The committee could not locate any information to indicate that either the Ag(II) process or the CerOx process had been developed significantly since 2001 in terms of their applicability to processing either stockpile or non-stockpile chemical agents or munitions.

Plasma Arc Technology

Plasma arc is a very high temperature process that has proved most effective for liquid waste streams. It could be used to destroy neat agent or treat secondary waste streams resulting from agent destruction, including detonation offgases. It could also be used for destroying metal parts, dunnage, and energetics.

Plasma arc technology utilizes the electrical discharge of a gas to produce a field of intense radiant energy and high-temperature ions and electrons that cause target chemical compounds to dissociate within a containment chamber. Large volumes of high- temperature vapor are generated that require a treatment system composed of a series of gas scrubbers, HEPA filters, and monitors to ensure that the system meets regulatory emission limits.

Variations of the plasma arc process are numerous and involve different plasma gases and reactor designs that provide either an oxidizing or a reducing environment. One system, developed by MGC Plasma AG in Switzerland (the MGC/PLASMOX process), has achieved destruction efficiencies greater than 99.99999 percent (seven nines) when processing adamsite, Clark I and II, phosgene, lewisite, yperite and a mixture of yperite, and lewisite. PLASMOX employs closely coupled, staged reaction zones (characterized as controlled pyrolysis) to completely destroy organic compounds. The Army has also investigated the PLASMOX process for destruction of neutralent waste streams as part of its technology test program (NRC, 2002).

Current Status

MGC/PLASMOX developed a portable unit, Model RIF 2, that was put into operation in 1994 and has since built additional units. The RIF 2 is skid-mounted and designed to be moved by four standard tractor-trailers. The unit has been used in Europe and is permitted under both Swiss and German environmental laws and regulations. It was used successfully to destroy chemical agents for the Swiss Army at its chemical materiel laboratory in Spiez, Switzerland. The PLASMOX tests run by the Germans and Swiss indicate that the system will destroy chemical agent safely and rapidly (Burns and Roe, 2001).

As part of a technology test program for non-stockpile CWM, the NSCMP hired Stone & Webster to conduct tests of the MGC/PLASMOX plasma arc process on simulated H and GB neutralents with MEA. MGC conducted these tests from January 8 through January 19, 2001, under a subcontract to Burns and Roe Enterprises at the MGC/PLASMOX facility in Switzerland.

The NSCMP had proposed that plasma arc technology be used primarily for the destruction of neutralent waste streams, e.g., it was a candidate for the direct destruction of the binary CWM components DF and QL, stored at Pine Bluff Arsenal. MGC/PLASMOX tests indicated a throughput rate for neutralent processing of approximately 13 liters per hour; process availability was reported to be 50 percent (NRC, 2002). However, there has been no recorded destruction of non-stockpile materiel by plasma arc technology in the United States.

MGC/PLASMOX technology was used successfully to destroy approximately 20 tons of toxic chemicals, including chemical munitions, in Albania. A portable unit was sent from Switzerland to Tirana, Albania, in time to begin the destruction of the toxic chemicals in July 2001. During the destruction process, unexpected major problems were encountered that affected the operation of the exhaust pipe, the air cooler, and the quenching process. In addition, the centrifuge in the plasma furnace was damaged by the high oxygen content in the destroyed pyrotechnic devices. Fortunately, these problems were solved without having a significant impact on the schedule. This project was completed in September 2001, and the plant was returned to Switzerland (Huber and Werner, 2002).

Technical Issues

The Army has identified approximately a dozen vendors of plasma arc technology in the United States, although none is currently permitted to treat hazardous waste or non-stockpile CWM (NRC, 2002).

Stone & Webster recommended that the MCG/PLASMOX system receive further testing on typical NSCMP liquid and solid waste streams, with particular attention paid to the deposition of solid materials in the system. Its report concluded that further improvements would have to be made to ensure that the system would comply with all EPA and state requirements (Stone & Webster, 2001).

Regulatory Approval and Permitting Issues

A number of regulatory issues were raised by the Army's test results for the MCG/PLASMOX technology that must be resolved before it could be permitted in the United States. These include improvements to the gas scrubber system, more complete knowledge of the fate of key components of the non-stockpile CWM (e.g., phosphorus), and better characterization of the solid, liquid, and gaseous secondary waste streams.

Public Concerns

The key public concern about plasma arc processes for the destruction of non-stockpile CWM in the United States centers on whether plasma arc offers a true alternative to incineration. Depending on the type of plasma gas used and the configuration of secondary oxidation zones, quench, and scrubber processes, plasma arc systems may produce gas volumes and reaction products that are quite similar to or quite different from those associated with incinerators. When oxygen is used as the plasma gas and/or if the plasma arc process has additional treatment chambers, the process may be practically indistinguishable from incineration. On the other hand, vendors often highlight the fact that plasma arc processes that do not use oxygen as the primary plasma gas differ from incineration, although even in these systems, oxidation generally takes place at a subsequent stage of the process. However, the levels of dioxins, furans, and other hazardous pollutants are likely to be below regulatory limits when the plasma arc system is optimally designed and controlled.

A noteworthy indication of the variance in public attitudes toward plasma arc is that in one case, after careful consideration, the Assembled Chemical Weapons Assessment program Dialogue Group accepted plasma arc as a valid alternative to incineration. However, in contrast with this, a spokesperson for the Non-Stockpile Chemical Weapons Citizens Coalition characterized plasma arc as a synonym for incineration and expressed concern that NSCMP was prematurely embracing the technology. As with incineration, the degree of public concern about plasma arc may vary with specific implementation and specific location (NRC, 2002).

Consequently, although the committee believes that this high-heat technology could successfully destroy the various chemical agents, its problem with public acceptance has kept it from being a viable option. Therefore, the committee determined that further support for the technology by the NSCMP would be a waste of resources and assigned this technology to Tier 2.

Photocatalytic Destruction System

The Scottish-based environmental technology firm Albagaia has developed two systems, one a laboratory-scale portable system and the other a trailer-mounted transportable system. These systems utilize photocatalytic technology for the destruction of chemical weapons agent. Photocatalytic technology is a heterogeneous electron transfer process (either oxidation or reduction) wherein a semiconductor such as titanium dioxide (TiO_2) is activated by light energy equal to or greater than the optical band gap of the semiconductor material, catalyzing the oxidation/reduction reaction. The organic agent is mineralized to carbon dioxide, water, and mineral acids or salts through the oxidation process. The photocatalytic process operates at near ambient temperature (±5°C) and pressure. In order for the photocatalytic process to be effective for the destruction of chemical munitions, the agent must be accessed and drained prior to treatment. The systems developed by Albagaia are not capable of accessing the agent nor are they effective for munitions hardware or energetics destruction.

The destruction of chemical contaminants by TiO_2 photocatalysis is well established. Laboratory, pilot, and field studies have demonstrated TiO_2-catalyzed photodegradation of a wide range of organic chemicals. Organic chemicals tested include aldehydes, alkanes, alkenes, amines, aromatics, carboxylic acids, dioxins, dyes, fuel constituents, halogenated hydrocarbons, herbicides, ketones, mercaptans, nitroglycerines, pesticides, polychlorinated biphenyls, solvents, and surfactants (Legrini et al., 1993). Typical concentrations of organic constituents that have been treated are on the order of 5 to 500 mg/L. The photocatalytic extraction of heavy metal (lead and mercury) contaminants from water has also been reported (Tennakone and Wijayantha, 1998).

Research efforts focused on materials relevant for the destruction of chemical weapons has been limited. An international collaboration between researchers in France, Russia, and the United States demonstrated complete mineralization of the mustard agent simulant in both air and water (Vorontsov et al., 2002).

The removal of arsenic via photocatalytic oxidation of arsenite to arsenate, followed by adsorption to the TiO_2 surface, has also been demonstrated (Bissen et al., 2001; Dutta et al., 2004).

Albagaia appears to be the first to develop commercial photocatalytic technology for the direct destruction of chemical weapons agent. It has developed two systems based on TiO_2 photocatalysis, the portable chemical agent destruction system (P-CADS) and the transportable chemical agent destruction system (T-CADS). These systems use a slurry form of TiO_2 powder in an aqueous batch system with an ultraviolet light source.

In the United States, a similar process was demonstrated on a rinsate generated during EDS operation. The Photo-Cat photocatalytic oxidation process was developed by Purifics Environmental Technologies of London, Ontario, Canada, for water treatment. The process employs ultraviolet light, hydrogen peroxide, and a titanium dioxide catalyst to destroy organic compounds. In 2001, the NSCMP tested the process at the Aberdeen Proving Ground (Burnham et al., 2002). The test solution was rinsate generated during the Rocky Mountain Arsenal bomblet destruction operation, in which the EDS was used to destroy four M139 bomblets with GB (sarin) fills. The test unit was rated at 7.2 kW, capable of treating 138 gallons of rinsate per day. Stable operation of the Photo-Cat system resulted in total organic carbon being reduced from 1,610 mg/L to 37.8 mg/L in 89 hours, a reduction of approximately 98 percent. Isopropyl methylphosphonic acid, a Schedule 2 compound, was reduced to below its detection limit of 3.94 mg/L from a starting concentration of 20.1 mg/L.

The technology was tested at Porton Down in 2003 with agent. Available results indicate a 99 percent destruction of HD within 6 hours using a system containing 6.42 m^2 of TiO_2 and continuous ultraviolet (UV) energy of 24 watts. The same system yielded 99.987 percent destruction of HD after 24 hours. To date, this is the only known photocatalytic system tested for chemical agent destruction.

While no known commercial applications of the photocatalytic technology are in use for chemical agent destruction, photocatalytic systems are in use for other applications. Systems have been permitted for use in the treatment of water and patented in the United States for the treatment of indoor air (Goswami, 1998).

The technology would need to be coupled with a munitions access and drainage system. To be considered for implementation, higher degrees of destruction would need to be demonstrated. The Albagaia systems are small enough to be easily transported. In addition they would be appropriately suited for situations where small amounts of agent are available at a given time, for a destroy-as-you-go system.

The photocatalytic process operates at near-ambient temperature and pressure and is inherently safe. The UV lamps pose the greatest hazard to worker safety, and these lamps are fully contained. In summary, photocatalytic technology uses TiO_2 and UV light for the mineralization of chemical agent. It is not capable of destroying munitions or energetics, and therefore must be coupled with another technology for complete destruction of munitions. The process operates at ambient conditions and does not require the use of reagents, and therefore is not likely to pose a hazard to workers or the environment during its operation. The products are not toxic and with the exception of arsenic would not require additional processing. Prior to adoption of this technology for agent destruction, additional testing for confirmation of efficacy for specific agents is required. While this technology is of interest for agent destruction, the low reaction rates involved and limited use to date precluded it from being considered as a Tier 1 technology.

Plasmazon

Plasmazon was developed in Germany and uses what are termed "activated ozone structures" for the destruction of chemical warfare materiel (Ehmer and Sieke, 1998). These activated ozone structures are characterized by unpaired or outer shell electrons and can be in one of three states: (1) basic (triplet or singlet) state $[O_3(^3B_2), O_3(^1B_2)]$; (2) excited state $[O_3(^1A_1,v)]$; or (3) ionized state (O_3^+, O_3^-). These species are generated using plasma ozone generation and withdrawing radicals from the discharge gap as quickly as possible after the dissociation of molecular ozone to atomic oxygen (reaction 1 below) and the generation of the ozone from reaction 2. The activated ozone structures have a half-life ranging from 70 msec to 70 sec, resulting in a higher level of radical generation than with the production and subsequent dissociation of ground state ozone.

$$O_2 \overset{hv}{\Longleftrightarrow} 2O \quad (1)$$

$$O_2 + O \Longleftrightarrow O_3 \quad (2)$$

Testing of the system for the destruction of Clark I has been reported (Sieke et al., 1998). The agent was diluted with

50 ml of acetone and the activated ozone passed through at a rate of 300 ml/min. After 1.5 hours of operation, a 98 percent reduction was reported. Multiple passes increased the reduction, with the fifth and final pass yielding a reported 99.999999 percent reduction from the initial concentration.

While the above results are encouraging, the committee determined that the Plasmazon technology had not yet developed sufficiently to justify including it in this report as a Tier 1 technology.

FINDING AND RECOMMENDATION

Finding 6-1. Based on currently available information, the Tier 2 technologies described in this chapter are not likely to meet foreseeable requirements of the NSCMP.

Recommendation 6-1. In the absence of significant developmental progress or unforeseen circumstances that would warrant reconsideration, the U.S. Army should not expend further resources on the evaluation of the following technologies for NSCMP applications:

- Acid digestion,
- Bulk vitrification,
- Firing pool,
- Biological approaches,
- DSTL electric furnace,
- Electrochemical oxidation,
- Photocatalysis,
- Plasma arc, and
- Plasmazon.

REFERENCES

Acid Digestion

Gaudre, M., M. Ferrari, and J.-M. Tauzia. 2001. Digestion process for securisation of recovered old ammunitions. Proceedings of the 2001 Chemical Weapons Demil Conference (CD). Porton Down, England: Defence Science and Technology Laboratory.

Soilleux, R. 2005. Possible Replacement Technologies for Operation Pongee After 2007. Porton Down, England: Defence Science and Technology Laboratory.

Bulk Vitrification

Buelt, J.L., C.L. Timmerman, K.H. Oma, V.F. Fitzpatrick, and J.G. Carter. 1987. In Situ Vitrification of Transuranic Waste: An Updated Systems Evaluation and Application Assessment. PNL-4800 Supplement 1, March. Richland, Wash.: Department of Energy.

Campbell, B., L. Thompson, and D. Fraser. 2005. Destruction of organic and explosive compounds using the GeoMelt® process. Available online at <http://www.dstl.gov.uk/news_events/conferences/cwd/2005/proceedings45.pdf>. Last accessed February 24, 2006.

CH2MHILL. 2005. Status and Direction of the Bulk Vitrification Program for the Supplemental Treatment of Low Activity Tank Waste at Hanford, RPP-24372-FP, Revision 0, January 12. Richland, Wash.: Department of Energy.

Loehr, C.A., J.R. Weidener, and S.O. Bates. 1992. Product Evaluation of In Situ Vitrification Engineering Test 4. EGG-WTD-9868, September. Idaho Falls, Idaho: Department of Energy.

Osborne, D. 2003. GeoMelt vitrification technologies. Proceedings of the 2003 Chemical Weapons Demil Conference (CD). Porton Down, England: Defence Science and Technology Laboratory.

Soilleux, R. 2005. Possible Replacement Technologies for Operation Pongee After 2007. Porton Down, England: Defence Science and Technology Laboratory.

Thomas, T.N., and R.L. Treat. 2001. Evaluation of In Situ Vitrification for Operable Unit 7-13/14. INEEL/EXT-2001-00279. Idaho Falls, Idaho: Department of Energy.

WDOE (Washington State Department of Ecology). 2005. DOE/ORP-2003-23, Rev. 1a. Available online at <http://www.ecy.wa.gov/programs/nwp/dbvs/sub2/enc5.pdf>. Last accessed February 28, 2006.

WNO (U.S. Water News Online). 2005. Hanford breaks ground on pilot project to glassify waste, February 2005. Available online at <http://www.uswaternews.com/archives/arcquality/5hanfbrea2.html>. Last accessed February 24, 2006.

Firing Pool

Guir, F. 1997. The technical challenge of dismantling and destroying old and abandoned chemical weapons. Pp. 156-165 in The Challenge of Old Chemical Munitions and Toxic Armament Wastes. T. Stock and K. Lohs, eds. Oxford, England: Oxford University Press.

Biological Approaches

Braatz, J.A. 1994. Biocompatible polyurethane-based hydrogels. Journal of Biomaterials Applications 9(1): 71-96.

Caldwell, S.R., and F.M. Raushel. 1991a. Detoxification of organophosphate pesticides using an immobilized phosphotriesterase from *Pseudomonas diminuta*. Biotechnology and Bioengineering 37(2): 103-109.

Caldwell, S.R., and F.M. Raushel. 1991b. Detoxification of organophosphate pesticides using a nylon based immobilized phosphotriesterase from *Pseudomonas diminuta*. Applied Biochemistry and Biotechnology 31(1): 59-73.

Gill, I., and A. Ballesteros. 1998. Encapsulation of biologicals with silicate, siloxane, and hybrid sol-gel polymers: An efficient and generic approach. Journal of the American Chemical Society 120(34): 8587-8598.

Gill, I., and A. Ballesteros. 2000a. Bioencapsulation within synthetic polymers (part 1): Sol-gel encapsulated biologicals. Trends in Biotechnology 18(7): 282-296.

Gill, I., and A. Ballesteros. 2000b. Bioencapsulation within synthetic polymers (part 2): Non-sol-gel protein-polymer composites. Trends in Biotechnology 18(11): 469-479.

Gill, I., and A. Ballesteros. 2000c. Degradation of organophosphorous nerve agents by enzyme-polymer nanocomposites: Efficient biocatalytic materials for personal protection and large-scale detoxification. Biotechnology and Bioengineereing 70(4): 400-410.

Grimsley, J.K., W.P. Singh, J.R. Wild, and A. Giletto. 2001. A novel, enzyme-based method for the wound-surface removal and decontamination of organophosphorous nerve agents. Pp. 35-49 in Bioactive Fibers and Polymers, ACS Symposium Series No. 742. J.V. Edwards and T.L. Vigo, eds. New York, N.Y.: Oxford University Press.

Havens, P.L., and H.F. Rase. 1991. Detoxification of organophosphate pesticide solutions: Immobilized enzyme systems. Pp. 261-281 in Emerging Technologies in Hazardous Waste Management II, ACS Symposium Series No. 468. D.W. Tedder and F.G. Pohland, eds. New York, N.Y.: Oxford University Press.

Kline, B.J., G. Drevon, and A.J. Russell. 2000. Enzymes in polymers and polymers from enzymes. Pp. 397-431 in Enzymes in Action, Green Solutions for Chemical Problems (NATO Science Series 1: Disarmament Technologies, Volume 33). B. Zwanenburg, M. Mikolajczyk, and P. Keilbanski, eds. Dordrecht, The Netherlands: Kluwer Academic Publishers.

Richins, R.D., A. Mulchandani, and W. Chen. 2000. Expression, immobilization, and enzymatic characterization of cellulose-binding domain-organophosphorus hydrolase fusion enzymes. Biotechnology and Bioengineering 69(6): 591-596.

Wang, A.A., A. Mulchandani, and W. Chen. 2002. Specific adhesion to cellulose and hydrolysis of organophosphate nerve agents by a genetically engineered Escherichia coli strain with a surface-expressed cellulose-binding domain and organophosphorus hydrolase. Applied and Environmental Microbiology 68(4): 1684-1689.

DSTL Electric Cylinder Furnace

Anderson, D.W., M.S. Sandford, and S. Kirkpatrick. 2003. Decomposition of sulfur mustard and sulfur mustard/lewisite mixture in an electric cylinder furnace. Proceedings of the 2003 Chemical Weapons Demil Conference (CD). Porton Down, England: Defence Science and Technology Laboratory.

Electrochemical Oxidation

NRC (National Research Council). 2001a. Disposal of Neutralent Wastes. Washington, D.C.: National Academy Press.

NRC. 2001b. Evaluation of Demonstration Test Results for Alternative Technologies for Demilitarization of Assembled Chemical Weapons: A Supplemental Review for Demonstration II. Washington, D.C.: National Academy Press.

Plasma Arc

Burns and Roe Enterprises. 2001. PLASMOX Technology for the Destruction of Chemical Warfare Materiel, March. Aberdeen Proving Ground, Md.: Program Manager for Chemical Demilitarization.

Huber, U., and D.B. Werner. 2002. VETOXA: Destruction of Toxic Chemicals in Albania, Report Number LS 2004-05, March 15. Spiez, Switzerland: Spiez Laboratory.

NRC (National Research Council). 2002. Systems and Technologies for the Treatment of Non-Stockpile Chemical Warfare Materiel. Washington, D.C.: The National Academies Press.

Stone & Webster. 2001. Evaluation of the Limited Engineering-Scale-Testing of the PLASMOX® Technology to Treat Chemical Warfare Materiel, July. Aberdeen Proving Ground, Md.: Program Manager for Chemical Demilitarization.

Photocatalytic Destruction

Bissen, M., M.M. Vieillard-Baron, A.J. Schindelin, and F.H. Frimmel. 2001. TiO_2-catalyzed photooxidation of arsenite to arsenate in aqueous samples. Chemosphere 44(4): 751-757.

Burnham, S.L., C.A. Freedlund, and J.M. Cardito. 2002. Evaluation of the Phase II Engineering Scale Testing of the Photo-Cat® Process to Treat NSCMP Rinsate Solutions, Revision 1, February. Aberdeen Proving Ground, Md.: Program Manager for Non-Stockpile Chemical Materiel.

Dutta, P.K., A.K. Ray, V.K. Sharma, and F.J. Millero. 2004. Adsorption of arsenate and arsenite on titanium dioxide suspensions. Journal of Colloid and Interface Science 278(2004): 270-275.

Goswami, D.Y. 1998. Goswami, D.Y., inventor. Universal Air Technology, assignee. November 10, 1998. Photocatalytic system for indoor air quality. U.S. Patent 5,835,840.

Legrini, O., E. Oliveros, and A.M. Braun. 1993. Photochemical processes for water-treatment. Chemical Reviews 93(2): 671-698.

Tennakone, K., and K.G.U. Wijayantha. 1998. Heavy-metal extraction from aqueous medium with an immobilized TiO_2 photocatalyst and a solid sacrificial agent. Journal of Photochemistry and Photobiology A: Chemistry 113(1): 89-92.

Vorontsov, A.V., A.A. Panchenko, E.N. Savinov, C. Lion, and P.G. Smirniotis. 2002. Photocatalytic degradation of 2-phenethyl-2-chloroethyl sulfide in liquid and gas phases. Environmental Science and Technology 36(23): 5261-5269.

Plasmazon

Ehmer, K., and R.W. Sieke. 1998. R.W. Sieke, inventor. May 14, 1998. Devices and methods for producing and using ozone. European patent WO9819961.

Sieke, R.W., G. Lippke, A. Krippendorf, R. Haas, and S. Lüdtke. 1998. Destruction of diphenylarsine chloride (CLARK I) with activated ozone. Environmental Science and Pollution Research 5(4): 199-201.

7

Assessing Large Burial Sites and Accessing Chemical Warfare Materiel

INTRODUCTION

The scope of work for the committee's study did not include a review of nontreatment technologies that might be associated with remote detection and accessing buried CWM. However, the U.S. Army requested the committee to compile any relevant technical and legal issues related to the need to detect, assess, access, and remediate the contents of large burial sites that were discovered while researching international destruction technologies.[1] This chapter briefly reviews these issues to provide some further background on technology considerations pertaining to large burial sites.

Large burial sites have not been thoroughly characterized, and their exact contents remain unknown. These sites may contain chemical ordnance of mixed types, fills, and condition, and miscellaneous debris including, in some cases, vehicles and other debris that were used at the sites for decontamination training. The chemical ordnance may be extremely deteriorated, especially in cases where the CWM was burned prior to burial.[2]

This scenario presents technical challenges for both assessing and accessing the CWM in these large burial sites.

ASSESSING LARGE CWM BURIAL SITES

DOD is a leader in the research and practice of detecting subsurface munitions and explosives of concern using geophysical processes. Since the mid-1980s there have been numerous investigative and remediation projects for conventional (high explosive) munitions and explosives of concern under various DOD programs such as the base realignment and closure program and the formerly used defense sites program.

Since that time, geophysical techniques and technologies for the detection of subsurface munitions and explosives of concern have been developed. It is now possible to detect with some accuracy individual or mass buried munitions and explosives of concern; magnetometry and active geophysical systems are the most common and productive technologies (ITRC, 2004).

In addition, DOD has programs supporting research and development in this technical area. Both the Environmental Security Technology Certification Program[3] and the Strategic Environmental Research Development Program[4] support research designed to improve this capability.

However, the technical challenges associated with assessing the contents of the identified large burial sites have not been specifically addressed. These technical challenges are caused by the intermingling of large buried masses of CWM with debris, which presents a complex geophysical signature. Although buried metal and metal masses are commonly detected using geophysical sensors, it is currently not possible to determine if a filled chemical munition is buried within a mass of metal debris using geophysical sensors. It is also not likely that this capability will be acquired in the near future,[5] and the committee's research into foreign technology did not reveal any potential breakthroughs in this area using geophysical sensors.

There are, however, some sensing technologies that should be investigated further. One is the use of chemical agent detector dogs to locate subsurface buried CWM. The committee was not able to review any literature in this area because all of the research was classified. However, it is

[1]William Brankowitz, PMNSCM, remarks at a meeting of the committee, September 7, 2005.

[2]William Brankowitz, PMNSCM, remarks at a meeting of the committee, November 29, 2005.

[3]See <http://www.estcp.org/index.cfm>.

[4]See <http://www.serdp.org/>.

[5]Meeting between Anne Andrews and Jeff Marquese, SERDP and ESTCP, and James Pastorick and Leonard Siegel, committee members, September 22, 2005.

known that the U.S. Bureau of Customs and Border Protection is using chemical detector dogs to detect CWM. The committee was informed that, although the research done to support this and other similar programs is classified, the chemical detector dogs have been demonstrated to have a detection capability "three to five orders of magnitude greater than the current best instrument detection capability."[6] The committee also found that England has plans to conduct tests at the Porton Down facility to determine the effectiveness of chemical agent detector dogs.[7]

If chemical agent detector dogs are demonstrated to be able to reliably detect CWM at very low concentrations, this capability could be applied to assessing large CWM burial sites. For example, it is to be expected that some of the sites, or portions of some of the sites, are free of chemical agent because no CWM was buried in that section of the burial or only empty CWM containers were buried there.[8] If chemical detector dogs could reliably confirm the absence of CWM, the excavation and removal of objects from portions of the burial pits so identified could possibly be carried out with reduced personal protective equipment and without other precautions normally taken for CWM excavation (negative pressure enclosures, for example).

There are also some potentially useful agent-sensing technologies that do not rely on biological sensors. These new devices may offer more rapid analysis and simpler, continuous measurement. One kind of new sensor is the electronic (or artificial) nose. An array of semiselective, cross-reactive sensors produces a response pattern characteristic of a chemical (Gardner and Bartlett, 1999; Albert and Walt, 2000). The patterns are preprogrammed mathematically so that upon exposure, the patterns are matched to the chemicals sensed. There are two main groups of electronic noses:

- Conducting sensor films and
- Optical sensor arrays.

The conducting sensor films are essentially thin films of materials that swell when exposed, causing resistance changes uniquely characteristic of a particular chemical. Since the process is reversible, these films can be used repeatedly. Their sensitivity has been demonstrated to be in the mid-ppb range for dimethyl methyl phosphonate and diisopropyl methyl phosphonate in air or exhaust fumes (Hopkins and Lewis, 2001).

The optical sensor arrays consist of etched optical fibers with attached dyes that change fluorescence depending on chemical polarity. The sensor's change in temporal fluorescence at a specific wavelength is monitored and matched to already determined patterns characteristic of known compounds. All of these arrays, electrical or optical, degrade with use.

Some of the new materials being developed that show promise for monitoring chemical agents include fluorescent indicator detectors, surface-enhanced Raman biosensing, and porous silicon technologies. A fluorescent indicator selective for electrophilic phosphates has been developed (Zhang and Swager, 2003). The use of structured nanoparticles coupled to surface-enhanced Raman-based biosensing makes it possible to reduce the size of a sensing unit substantially (Yonzon et al., 2004; Shafer-Peltier et al., 2003).

Based on the optical thickness of films on porous silicon, this new generation of sensors relies on changes in the film to detect chemicals. In tests for volatile organic compounds, polycyclic aromatic hydrocarbons, explosives, and other chemicals, these sensors have been sensitive to ppb ranges (Sailor, 1997).

Lab-on-a-chip technology is advancing rapidly because it has so many potential applications, is small enough for field use, and fast. There are several microchip protocols for monitoring chemical agents and their degradation products, including a precolumn enzymatic reaction, a capillary electrophoresis/conductivity microfluidic device, and a capillary electrophoresis microchip separation and amperometric detection device (Wang et al., 2002, 2004a, 2004b). None of these technologies has been tested extensively enough to allow recommending their use, but they do have the potential to improve current agent-sensing capabilities due to their small size, low power requirements, lower cost, and increased speed.

However, the only reliable method of identifying the contents of a mixed CWM and debris-filled burial pit using currently available known technologies, or technologies likely to be available within the next 5 to 10 years, is archeological excavation—that is, carefully excavating the overburden and accessing the contents for visual identification and nondestructive testing. This nothwithstanding, there appear to be significant possibilities for technology transfer. It may be cost-effective for DOD (including the U.S. Army) to coordinate with other U.S. government agencies to evaluate results from ongoing research programs.

Finding 7-1a. A critical factor in ensuring buried CWM are adequately addressed is developing cost-effective, reliable methods of detecting the presence of buried CWM remotely.

Finding 7-1b. Several U.S. government agencies are investigating remote sensing techniques to detect chemical agents, non-CWM munitions, and buried hazardous waste. Some of this research may be applicable to detecting and assessing buried CWM.

---

[6]Personal communication between Jerry Walsh, Department of Homeland Security, and Harrison Pannella, NRC staff, November 16, 2005.

[7]Personal communication between Richard Sollieux, DSTL Porton Down, England, MoD, and Richard Ayen, committee chair, January 13, 2006.

[8]William Brankowitz, PMNSCM, remarks at a meeting of the committee, November 29, 2005.

Recommendation 7-1. The U.S. Army should coordinate with other federal agencies on developing an easy-to-use, comprehensive database and on the evaluation of remote techniques to detect buried CWM in a reliable but cost-effective manner.

ACCESSING THE CONTENTS OF LARGE BURIAL SITES

Accessing Techniques in Other Countries

The committee's research into foreign technologies showed that almost all foreign countries use a low-tech approach—manual excavation—to accessing buried CWM.[9] One exception to this is Japan, which is currently planning a combination telerobotic and automated CWM excavation and handling system for the large burial site at Haerbaling, Jinlin Province, China, to dispose of chemical weapons abandoned by the Japanese. This site consists of two very large burial pits expected to contain between 300,000 and 400,000 individual chemical munitions.

For this project, the Japanese are designing a remotely operated and automated excavation system consisting of excavation robots, a device to remove attached soil using pressurized air, and an automated transportation system that will take the removed CWM through a series of cleaning and assessment stations and then finally to a packing station and temporary storage.[10]

Although it is not possible for the committee to evaluate a system that has not yet been designed, the concept of automated or telerobotic excavation and handling of CWM deserves to be evaluated.

It is acknowledged that use of robotic systems for excavation and handling of CWM is likely to result in less delicate handling of the CWM than is possible using trained hazardous materials technicians to perform these tasks. This can be seen in the Japanese design for the Haerbaling system, which assumes the unplanned detonation of one out of every 1,000 CWM handled.[11] This risk may be unavoidable due to the deteriorated condition of the explosively configured munitions. Furthermore, it may be acceptable if the system is designed to handle the unplanned detonations without serious equipment damage or the release of chemical agent.

What is instructive about the planning of this system is that it demonstrates the trade-offs between preventing unplanned detonations and surviving unplanned detonations, which should be evaluated before deciding on the approach that will be used to excavate and handle CWM at the large U.S. burial sites. The trade-offs include these:

- Using trained technicians to manually remove and handle the CWM is likely to result in more delicate handling of the CWM and fewer unplanned detonations. However, an unplanned detonation in this scenario is likely to have catastrophic and unacceptable consequences (severe injury or death of the technicians).
- Using telerobotic or automated robotics to perform the excavation, removal, and handling of CWM will result in rougher handling and more unplanned detonations. However, adequate engineering of the system will result in less serious consequences for each unplanned detonation.

The current technology for robotics is mature, and a significant amount of research and development is being done by private companies to support advances in manufacturing processes.[12] Much of this technology is applicable to the development of robotic systems for use on large CWM burial sites.[13]

As a result, the best solution to accessing CWM in large burial sites may be a combination of manual removal using trained technicians and, when the risk is unacceptable, removal by a suite of mobile robotic systems specially developed to perform specific high-hazard tasks as needed.

Processes for Close Proximity and In Situ Treatment

The likelihood that large CWM will be found in a deteriorated condition means that the Non-Stockpile Chemical Materiel Project (NSCMP) must be able to treat large CWM in place without unnecessary movement. This is true for burial sites containing large numbers of CWM as well as for sites containing few or even single items. Research by the committee into foreign solutions to this problem indicates that no new foreign technology has been developed or is likely to be developed that is useful to NSCMP.

Most other countries respond to this scenario by open detonation of the unstable CWM using at least 5 pounds of explosive for each pound of chemical agent.[14] In this situation, it is hoped that the high detonation temperature of the donor explosive will consume a large portion of the chemical fill. Some countries employ considerably larger amounts of explosive in an attempt to maximize the destruction of the agent fill. However, such a solution is considered by the

[9]Teleconference between Jeffrey Osborne, OPCW, and the committee, December 9, 2005.

[10]Personal communication between Takayuki Matsuda, Deputy Director, Abandoned Chemical Weapons Cabinet Office, Government of Japan, and Douglas Medville, committee member, December 9, 2005.

[11]Teleconference between Jeffrey Osborne, OPCW, and the committee, December 9, 2005.

[12]See <http://telerobotics.stanford.edu/publications.htm>; <http://brl.ee.washington.edu/Publications/Publications_Index/All_Reports_Index.html>; and <http://www.ri.cmu.edu/cgi-bin/tech_reports.cgi>.

[13]See <http://www.army-technology.com/contractors/mines/telerob/> and <http://www.foster-miller.com/lemming.htm> for currently available explosive ordnance disposal robots.

[14]Teleconference between Jeffrey Osborne, OPCW, and the committee, December 9, 2005.

committee to be generally unacceptable in the United States owing to the difficulty of obtaining regulatory approval for this method of disposal.

Large Item Transportable Access and Neutralization System/Monica™

The committee's research has shown that the NSCMP is a leader in developing technology to address this scenario. The NSCMP has already developed the system requirements for the Large Item Transportable Access and Neutralization System (LITANS) and is currently developing the system (U.S. Army, 2005).

The requirements call for a system that can house and contain a CWM up to a ton container in size. Moreover, the system will be portable and allow for drilling, sampling, agent removal, and neutralization while preventing releases of agent during processing.

The NSCMP is also currently using the Monica remote case entry and sampling system manufactured by MMIC EOD of England.[15] This is a commercially available remote annular drill and seal system with a vacuum mounting system. According to the NSCMP, this system has been determined to be useful for accessing and removing agent fills from large CWM. However, some leakage has occurred, which prevents use of this system without vapor containment.[16] This requirement notwithstanding, according to the NSCMP, it is likely that the Monica remote case entry and sampling system will be an acceptable solution to accessing the agent in large CWM as long is it is used within a containment structure such as LITANS.

All of the above systems, however, require that the CWM be moved at least a small distance into the LITANS enclosure. This may not be acceptable in the case of an extremely deteriorated or possibly shock-sensitive CWM that has been determined to be unsafe to move. In this case an alternate containment system that can be installed over the CWM at its existing location is needed.

Ballistic Tent-and-Foam System for Vapor Containment

One potential solution to dealing with extremely deteriorated or shock-sensitive CWM has been investigated by the NSCMP: the ballistic tent-and-foam system. According to the testing plan for the ballistic tent-and-foam system,

> The testing would involve a field tent/foam system to suppress the blast overpressure and stop the fragments from a simulated chemical munitions scenario where the munition cannot be moved and must be blown in place. The system to be tested would involve a 2-tent system, an inner and an outer tent.
>
> The outer tent, measuring approximately 13 ft × 10 ft × 8 ft tall, is placed over the munition. The inner tent, 7 ft in diameter at the base and tapering to 4 ft at the top, will be placed inside the outer tent and directly over the explosive device. Neither of the tents have a floor. The inner tent will then be filled with a Silvex soap-based foam formulation.
>
> The foam also contains decon solution. . . . Then the secondary tent is placed over the primary and is hooked up to the Air Pollution Control (APC) equipment. The scrubber, the first piece of the APC, will contain a NaOH solution. After use, the waste liquid will be disposed of and will consist of the NaOH, some Oil of Wintergreen Residue [Methyl salicylate, an agent simulant used in testing], some explosive residue, and decon agents in the foam.
>
> Past studies have shown that this aqueous foam is a good material to suppress the blast from an explosion. The main role of the tent system is to stop or reduce the fragmentation that occurs in a detonation. (U.S. Army, Undated)

The government of England is currently using a similar double-tent containment system for in-place disposal of CWM by detonation.[17]

It is the opinion of the committee that the LITANS/ Monica system is the most promising solution for the disposal of large CWM and ton containers that can be moved into the LITANS containment system. However, there is a need for a system to allow in-place disposal of deteriorated and unstable CWM without moving the munition.

It is possible that a hybrid application of the Monica remote case entry and sampling system and a tent-like containment structure and APC system, similar to the existing ballistic tent used for the tent-and-foam system, can be easily developed, tested, and fielded to fill this need. This new system would use the tent-and-foam and APC technology to contain and capture any chemical agent released during the agent removal via the Monica without requiring the munition to be moved from the location where it was found.

According to PMNSCM, the results of its testing of the tent-and-foam detonation system have not been promising, and the testing of this system has been discontinued.[18] This leaves a gap in the CWM disposal capabilities of PNMSCM because there is currently no method for in-place disposal of small CWM by detonation that is acceptable to most environmental regulators.

REFERENCES

Albert, K.J., and D.R. Walt. 2000. High-speed fluorescence detection of explosive-like vapors. Analytical Chemistry 72(9): 1947-1955.

[15] See <http://www.mmic-eod.co.uk/Equipment%20Page/Equipment/Monica/monica%20page.htm>.

[16] William Brankowitz, PMNSCM, remarks at a meeting of the committee, November 29, 2005.

[17] Personal communication between Richard Sollieux, DSTL Porton Down, England, MoD, and Richard Ayen, committee chair, January 13, 2006.

[18] Personal communication between Laurence Gottschalk, PMNSCM, and Harrison Pannella, May 10, 2006.

Gardner, J.W., and P.N. Bartlett. 1999. Electronic Noses: Principles and Applications. Oxford, England: Oxford University Press.

Hopkins, A.R., and N.S. Lewis. 2001. Detection and classification characteristics of arrays of carbon black/organic polymer composite chemiresistive vapor detectors for the nerve agent stimulant dimethylmethylphosphonate and diisopropylmethylphosponate. Analytical Chemistry 73(5): 884-892.

ITRC (Interstate Technology Regulatory Council). 2004. Geophysical Prove-Outs for Munitions Response Projects, November. Washington, D.C.: Interstate Technology Regulatory Council Unexploded Ordnance Team.

Sailor, M.J. 1997. Sensor applications of porous silicon. Pp. 364-370 in Properties of Porous Silicon. L. Canham, ed. Exeter, England: Short Run Press Ltd.

Shafer-Peltier, K.E., C.L. Haynes, M.R. Glucksberg, and R.P. Van Duyne. 2003. Toward a glucose biosensor based on surface-enhanced Raman scattering. Journal of the American Chemical Society 125(2): 588-593.

U.S. Army. 2005. System Requirements Document for the Large Item Transportable Access and Neutralization System, October. Aberdeen Proving Ground, Md.: Program Manager for the Elimination of Chemical Weapons.

U.S. Army. Undated. Environmental Assessment for Ballistics Testing of Tent and Foam for Use in Removal Actions at Aberdeen Proving Ground, Maryland. Available online at <http://www.apg.army.mil/apghome/sites/directorates/restor/environmental_assessment.htm>. Last accessed May 3, 2006.

Wang, J., M. Pumera, G.E. Collins, and A. Mulchandani. 2002. Measurements of chemical warfare agent degradation products using an electrophoresis microchip with contactless conductivity detector. Analytical Chemistry 74(23): 6121-6125.

Wang, J., G. Chen, A. Muck, Jr., M.P. Chatrathi, A. Mulchandani, and W. Chen. 2004a. Microchip enzymatic assay of organophosphate nerve agents. Analytica Chimica Acta 505(2): 183-187.

Wang, J., J. Zima, N.S. Lawrence, M.P. Chatrathi, A. Mulchandani, and G.E. Collins. 2004b. Microchip capillary electrophoresis with electrochemical detection of thiol-containing degradation products of V-type nerve agents. Analytical Chemistry 76(16): 4721-4726.

Yonzon, C.R., C.L. Haynes, X. Zhang, J.T. Walsh, Jr., and R.P. Van Duyne. 2004. A glucose biosensor based on surface-enhanced Raman scattering: Improved partition layer, temporal stability, reversibility, and resistance to serum protein interference. Analytical Chemistry 76(1): 78-85.

Zhang, S.W., and T.M. Swager. 2003. Fluorescent detection of chemical warfare agents: Functional group specific ratiometric chemosensors. Journal of the American Chemical Society 125(12): 3420-3421.

Appendix A

Tables Illustrative of a Variety of Non-Stockpile Items

Tables A-1 through A-4, reprinted from an earlier report (NRC, 2005), show inventories of non-stockpile items that have been stored at various military sites for eventual treatment and disposal. The tables are reproduced here to illustrate the variety of non-stockpile items that exist and that could be encountered during future recovery operations at non-stockpile burial sites.

REFERENCES

NRC. 2005. Impact of Revised Airborne Exposure Limits on Non-Stockpile Chemical Materiel Program Activities. Washington, D.C.: The National Academies Press.

Verrill, D., and J. Salcedo. 2001. X300P90 Characterization Project. Preliminary Data Review, May 14. Aberdeen Proving Ground, Md.: Product Manager for Non-Stockpile Chemical Materiel.

TABLE A-1 Inventory of Non-Stockpile Items at the Pine Bluff Arsenal

		No. Containing a Chemical(s)									
Item	No. Empty	H/HD/HN/ HS/HT	GA/GB/ GD	VX	DM/L	CG/CK	DF	QL	Other	Unknown	Total No. of Items
Munition											
4.2-in. mortar round	596[a]	99[a]				1[b]				36[a]	732[a]
75-mm projectile	4[a]	9[a]									13[a]
200-mm Livens projectile	3[a]	5[a]				3[b]					11[a]
4.7-in. projectile		1[a]									1[a]
155-mm projectile	1[a]										1[b]
105-mm projectile										1[a]	1[b]
M70A1 bomb (poss. explosive)	6[a]	3[a]									9[b]
150-mm German Traktor rocket w/expended motor	224[a]	184[a]									408[a]
150-mm German Traktor rocket w/unexpended motor	13[a]	18[a]									31[a]
150-mm German Traktor rocket w/warhead only	26[a]	12[a]									38[a]
Subtotal	873[a]	331[a]				4[b]				37[a]	1,245[a]
Chemical sample container[c]											
Ton container		2[d]	2[d]								4[d]
4-in. cylinder		2[d]									2[d]
Lab sample container			14[d]	7[d]							21[d]
Vial (L)					1[b]						1[b]
Subtotal		4[d]	16[d]	7[d]	1[b]						28[d]
Chemical agent ID set (CAIS)											
Mustard (H/HD/HS)		5,764[b]									5,764[b]
Nitrogen mustard (HN-1 and -3)		50[b]									50[b]
Lewisite (L)					397[b]						397[b]
Chloropicrin (PS)									396[b]		396[b]
Phosgene (CG)						396[b]					396[b]
Chloroacetophenone (CN)									17[b]		17
Adamsite (DM)					17[b]						17[b]
Triphosgene (TP)									17[b]		17[b]
Cyanogen chloride (CK)						33[b]					33[b]
Diethyl malonate, etc. (GS)									33[b]		33[b]
Subtotal		5,814[b]			414[b]	429[b]			463[b]		7,120[b]
Binary agent precursor											
M20							56,764[d]				56,764[d]
Drum							7[d]	291[d]			298[d]
Subtotal							56,771[d]	291[d]			57,062[d]
Empty ton container[e]					4,375[b]						4,375[b]
Total	873[a]	6,146[a]	2[b]	2[b]	4,789[b]	433[b]	56,771[d]	291[d]	463[b]	37[a]	69,830[d]

[a]Data from Verrill and Salcedo (2001).

[b]Provided to the Committee on Review and Evaluation of the Army Non-Stockpile Chemical Materiel Disposal Program by the PMNSCM on July 10, 2001.

[c]Inventory consists of individual CAIS items, not complete CAIS.

[d]Provided to the committee by Darryl Palmer, Office of the PMNSCM, on February 14, 2005.

[e]Sampling of some of these containers indicated that they may be contaminated with lewisite, arsenic, and/or mercury.

SOURCE: Reprinted from NRC, 2005.

TABLE A-2 Inventory of Non-Stockpile Items at Dugway Proving Ground (DPG) and Deseret Chemical Depot (DCD), Utah[a]

		Chemical Fill				
Item	Location	H/HD/HN/HT/HS	GA/GB/GD	Lewisite	VX	Total No. of Items
Explosive munitions (4.2-in.mortar rounds)	DPG	8[a]				
Chemical sample containers						
Ton container	DCD	1				1
Containers, bottles, vials	DPG				18[b]	18
Containers (39 HD, 5 HT)	DCD	45				45
Ampoule	DCD		1			1
Total		54	1	1	18	65

[a]Provided to the committee by PMNSCM on November 19, 2004.
[b]Scheduled for transport to DCD.

SOURCE: Reprinted from NRC, 2005.

TABLE A-3 Inventory of Non-Stockpile Items at Aberdeen Proving Ground, Maryland[a]

	Chemical Fill					
Item	HD/HT/HS	GB/GA/GD	VX	Lewisite	CG	Total
Chemical sample containers						
55-gallon drums (pumpkins)	10					10
30-gallon buckets (pumpkins)		5	5			10
5-pint cans (vials or bottles)	3		16			19
Steel cylinders		12				12
Multipack bottles, vials		8		9		17
DOT bottle			1			1
Ton container		1				1
Total	13	26	22	9		70

[a]Provided to the committee by PMNSCM on November 19, 2004.

SOURCE: Reprinted from NRC, 2005.

TABLE A-4 Inventory of Non-Stockpile Items at Anniston Chemical Activity, Alabama[a]

	Chemical Fill			
Item	HD/HT	GB	VX	Total
Chemical sample containers				
Vials		36		36
DOT bottles	5		7	12
Ton containers		2		2
Total	5	38	7	50

[a]Provided to the committee by PMNSCM on November 19, 2004.

SOURCE: Reprinted from NRC, 2005.

Appendix B

Tier 1 Munitions Processing Evaluation Subfactor Comparative Tables

TABLE B-1 Process Maturity Subfactor Evaluations for Tier 1 Munitions Processing Technologies

	Tier 1 International Munitions Processing Technologies			
Maturity Subfactors	CDC	DAVINCH	Dynasafe	U.S. EDS
Is the technology in use for any type of material, even one not related to CWM in the U.S.?	Yes. Has been used extensively to destroy conventional munitions including smokes, WP, and CS.	No.	Dynasafe explosive containment chambers are in use in the U.S., but chemical munition processing and agent destruction has not been demonstrated in the United States.	Yes, the EDS has been used extensively to destroy a variety of old munitions and agents, including GB nerve agent.
Is the technology in use for any type of material, even non-CWM related internationally?	Yes. Used in Belgium to destroy over 2,000 projectiles containing Clark II agent.	Yes. Used in Japan to destroy 500 bombs containing Clark I and II and 100 bombs containing a 50:50 mix of L and H.	Yes. Static kilns have been used in several countries to destroy conventional munitions.	Yes, it has destroyed several types of munition in testing at Porton Down, U.K.
Has the technology been permitted or otherwise approved in the U.S. for CWM or energetics?	Yes. Permitted to destroy high explosive munitions.	No.	Yes, for energetics (e.g., the burster detonation vessel at MAPS). No for chemical munitions.	Yes.
Has the technology been permitted or otherwise approved in the U.S. for industrial wastes?	No, not intended for use with such wastes.	No.	No, not intended for use with such wastes.	N/A
How much, if any, additional R&D is required in order to implement the technology?	Little, if any. Process appears to be mature. Testing may be required to demonstrate productivity and ability to destroy nerve agents.	None. Process appears to be mature. Testing may be required to demonstrate ability to access agents in U.S. non-stockpile munitions and to destroy nerve agents.	Proven technology for destroying conventional munitions containing energetics. Additional R&D not needed for this purpose. Testing required to demonstrate ability to destroy chemical warfare materiel.	None.
What, if any, are the scale-up requirements needed to implement the technology?	None. Available models can destroy projectiles up to 210 mm in diameter.	None. DAVINCH has destroyed large Japanese recovered CWM (1 meter long, 0.2 meters diameter, 19 kg mustard agent/lewisite agent mix). Volume of inner vessel is 30 times that of EDS-2 and explosive containment is 20 times EDS-2.	None. May want to increase explosive containment capability beyond 5.1-lb TNT equivalent or increase physical size of detonation chamber beyond 2-meter diameter if need exists for greater capability.	None, although there are size limitations on the types of munitions that can be destroyed.
Can the technology be implemented within 3 to 5 years?	Yes. Already has been implemented in Belgium.	Yes. Already has been implemented in Japan.	Yes. Being implemented at German government chemical weapon destruction facility.	Yes, in full use now.
Overall	Employed routinely in Belgium for recovered CWM; tested extensively at Porton Down.	Mature process has been used but only for one application in Japan. Very well engineered; designed specifically for chemical munition destruction.	Mature, well-engineered process for conventional munitions; has demonstrated ability to destroy agent and is acquiring experience with chemical weapons, both inert and containing explosives.	Fully mature.

TABLE B-2 Process Efficacy/Throughput Subfactor Evaluations for Tier 1 Munitions Processing Technologies

	Tier 1 International Munitions Processing Technologies			
Process Efficacy Subfactors	CDC	DAVINCH	Dynasafe	U.S. EDS
What is the DRE?	DRE not published, but likely exceeds six 9s. In U.S./U.K. tests with current operating procedures, destruction in detonation chamber is 99.99 percent.	99.9999 percent claimed for agent removal in primary destruction in vessel. 99.9 percent claimed removal of any remaining agent in offgas in secondary treatment. 99.99 percent removal of arsenic from metal surfaces and dust also is claimed.	Claimed to be 99.9999 percent. Need test data from past agent destruction and operating results from munition destruction in Germany for confirmation.	DRE after neutralization step is 99.9999 percent (six 9s) or better.
Does agent destruction meet the terms of the CWC (irreversible and verifiable)?	Irreversible: yes. Verifiable: yes.	Irreversible: yes. Verifiable: yes.	Yes, if agent in munition bodies can be accessed and destroyed. Need to be able to confirm agent destruction if deflagration occurs, agent escapes through weak point in the munition, and munition emerges intact from the detonation chamber.	Yes, irreversible and verifiable.
What is the DRE for energetics?	DRE not published, but should approach 100 percent under confined detonation conditions.	Claimed to be 100 percent in fireball in vessel. Test results not seen but no reason to doubt nearly complete energetics destruction under temperature and overpressure conditions in vessel.	Claimed to be 100 percent; no reason to doubt nearly complete destruction given high temperature in the detonation chamber.	Essentially complete.
Is the process reliable and robust?	Appears to be both. Cause of two misfires in an early test has been corrected.	Appears to be both. Successful destruction of 600 chemical bombs without needing to replace inner vessel. All munitions and agent appear to have been destroyed on first attempt, but independent verification needed.	Very robust and reliable for conventional munitions; being demonstrated for chemical munitions.	Yes, has performed reliably under field conditions with a variety of agents, munition types, degrees of corrosion.
Is the process highly complex or relatively simple?	Basic process is simple. Software and hardware for automated handling of munitions and for control of emissions add complexity.	Moderately complex operation involving placement of slurry explosive around munition, use of donor charges, and robotic handling to bring munitions into vessel. Agent and energetics are destroyed in a three-step process. Secondary treatment of metal parts is not required. Offgas treatment appears to be straightforward but reprocessing of offgas may be needed if agent is detected prior to release.	Front end is simple (thermal destruction of munition energetics and agent) although some material handling is involved. Back end for offgas processing can be complex—e.g., particulate removal, combustion, offgas scrubbing, and filtration; depending on agent being processed and regulatory requirements for offgas treatment.	Relatively simple, but requires multiple operational steps.

continued

TABLE B-2 Continued

	Tier 1 International Munitions Processing Technologies			
Process Efficacy Subfactors	CDC	DAVINCH	Dynasafe	U.S. EDS
What are the personnel/ staffing requirements for the technology?	Usually 18 staff needed for deployment of larger models.	Total staff requirements appear to be 20-25 people. Number of operators and supervisory personnel not known.	Staff needed to operate Dynasafe: 2-4 people plus control room and support staff. Total staff requirements should be 4-8 people.	EDS operators, air monitoring system operators, and analytical support staff.
What is the process throughput?	Potentially 40 projectiles per shift in T-60 model, but needs to be demonstrated. Five munitions per hour achieved in Belgian tests.	Moderate throughput; 3 shots/day with 3 Yellow bombs/shot demonstrated; 5 shots/day with 3 Yellow bombs/shot claimed. Also, 5 shots/day with 8 75-mm and 90-mm shells = 40 shells/day are claimed but not demonstrated.	Throughput varies greatly with the munitions and how the Dynasafe is operated. Estimated to be as low as 20 items/ 10-hour day for large munitions, e.g., 8-in. projectile and as high as 480/10-hour day for small items, e.g., M139 bomblets.	One detonation per 2 days, but each detonation may destroy three small weapons (EDS-) or up to six (EDS-2).
Is the process scalable so that it can address small, medium, and large munition finds?	Yes. Three sizes appear to provide appropriate capability for small, medium, and large quantities. All are transportable to the site.	Yes. There are two DAVINCH sizes available and multiple units can be deployed. Large size (70 tons) of smaller DAVINCH (DV45) may make it impractical for small quantity finds, however.	Yes. Three detonation chamber sizes are available and several units can be used together to meet throughput requirements.	The EDS seems appropriate for small or medium caches. At Pine Bluff Non-Stockpile Facility, several EDSs could have destroyed ca. 1,000 munitions in an acceptable time period.
Is the process capable of handling multiple munition types?	Yes. Has handled chemical projectiles up to 155-mm. Has destroyed wide range of conventional munitions.	Has handled only two types of Japanese bombs so far. Needs to demonstrate ability to destroy other munition types.	Yes. Has been used for a large variety of conventional items, e.g., grenades, cartridges, mortar rounds, projectiles, mines, fuzes.	Yes, but size-limited.
Is the process capable of handling multiple agent types?	Has destroyed CG, PS, blister agents, and riot control agents but not nerve agents.	Has destroyed vomiting agents and blister agent mix (mustard agent/lewisite) but not nerve or other agents.	Will be demonstrated during operations with German chemical munitions (e.g., mustard agent, lewisite, Clark agent).	Yes.
Is the process transportable?	Three CDC models can be moved on flatbed trucks (1 to 8 trailers depending on model).	A DAVINCH that can be moved on a flatbed trailer is under development but not yet built.	Yes. A mobile version of the SK1200 can be transported in eight containers on three flatbed trailers.	Yes.
Overall	Effective for several munition and agent types; not tested for nerve agents.	Well-engineered technology designed specifically for chemical munition destruction. Has performed reliably and effectively, but needs more operational experience with additional munition and agent types.	Well-engineered unit that is reliable and effective for many conventional munitions and ammunition. Is expected to be effective in destroying chemical munitions present in Germany and has destroyed mustard-filled grenades.	Broadly efficacious.

TABLE B-3 Process Safety Subfactor Evaluations for Tier 1 Munitions Processing Technologies

	Tier 1 International Munitions Processing Technologies			
Process Safety Subfactors	CDC	DAVINCH	Dynasafe	U.S. EDS
What are the worker safety and health risks?	Low risk. Most operations are conducted remotely other than encasing the munition in precast explosive and attaching and detaching the firing wires. Safety interlocks minimize risk of premature explosion of "donor" explosive. Workers are in level C PPE except during maintenance, when level B is worn.	Low risk. All operations are conducted remotely other than munition loading in boxes and injection of slurry explosives around munition. Workers are in level D PPE.	Low. Level D PPE used for munition loading. Once munitions are in boxes, process is automated with no worker exposure under normal conditions.	Primarily manual handling of a munition during preparation and loading.
What are the community safety and health risks?	Apparently low. Detonation chamber is double-walled. Improvements in ventilation and offgas cleanup minimize potential emissions. Solid wastes reported suitable for transport under government control.	Apparently low. DAVINCH is double-walled pressure vessel inside a larger structure.	Low to none. Munition destruction is in double-walled detonation chamber in an external shell. Technology has track record of destroying conventional munitions safely. Ability to contain offgases following munition and agent destruction expected to be demonstrated during operation in Germany.	Minimal. Possible agent release during handling of a weapon before loading is complete.
What are the process monitoring requirements?	Agent monitors (MINICAMS) are located before and after the carbon filter beds before releasing offgas to the atmosphere.	Agent monitors (MINICAMS) are used to test offgases for residual agent; gas can be returned to vessel for additional destruction if needed. Monitors are also located between the two carbon filter beds before releasing combustion gas to atmosphere.	In operation at Munster, Germany, very few. Design philosophy is that all hardware provides total containment and that all offgases are extensively treated; thus extensive monitoring using MINICAMS or other hardware is not necessary.	The contents of the chamber/reactor are monitored before liquid and vapor are released.
To what extent have engineering controls been developed to ensure process safety?	Engineering controls (hardware, software, safety interlocks) have been added to control automated loading, firing, and product gas cleanup.	Extensive design and testing conducted on the DAVINCH vessels; monitoring of fatigue damage and crack initiation; remote operation using robotic arm to transport munition into vessel; very heavy dual-walled construction of vessel and door.	Extensive engineering controls, process monitoring, and data acquisition are used.	Steel plates surrounding munition protect the 2-in. thick stainless-steel vessel which operates in a vapor containment structure. Detonation and reagent introduction are conducted remotely. Air and liquid effluents are monitored for presence of agent.
Overall	Appears to be a safe process for workers and the public.	Appears to be a safe process for workers and the public.	Appears to be a safe process for workers and the public.	Appears to be a safe process for workers and the public.

TABLE B-4 Public and Regulatory Acceptability in a U.S. Context Subfactor Evaluations for Tier 1 Munitions Processing Technologies

	Tier 1 International Munitions Processing Technologies			
Public and Regulatory Acceptability in a U.S. Context Subfactors	CDC	DAVINCH	Dynasafe	U.S. EDS
Is the process inherently incineration-like?	Agent destruction by fireball and shock wave in vessel. Offgas treatment involves catalytic oxidation.	Agent destruction by fireball and shock wave in vessel, but offgas treatment in combustion chamber is incineration. Use of a cold plasma to treat offgas is less incineration-like since gas molecules are being ionized rather than combusted.	Thermal destruction of agent in detonation chamber may be considered to be incineration-like, although heating is indirect and there is no contact with a flame. Offgas treatment in combustion chamber is incineration.	No.
Does the process break key chemical bonds (e.g., C-P bond for nerve agents)?	Yes, for various agents tested. Has not processed nerve agents.	Yes, for agents tested: mustard agent, lewisite, vomiting agents. Has not processed nerve agents.	Yes, for agents tested (vomiting agents, As Cl_3, mustard agent). Not known for nerve agents.	Yes, those bonds related to toxicity that are not broken during detonation are broken during neutralization.
Could the process produce dioxins or other notable by-products?	Proponent says no, but independent confirmation is needed.	Proponent says no, but independent confirmation is needed.	Not known. Depends on offgas treatment.	Not detectably.
Does the process allow holding and testing of process residuals prior to release?	No. Offgases go through adsorption, oxidation, and carbon filtration but are not held for retreatment.	Yes. Offgases go to a holding tank for storage, testing, and re-treatment if necessary. Offgas can be retreated in either the DAVINCH vessel or in a cold plasma unit to ensure agent destruction.	Not when operated as an open system. If operated as a closed system, detonation offgas can be held in detonation chamber, tested for agent, and heated until agent in gas is reduced to acceptable level.	Yes.
Does the process result in excessive noise, odors, or other nuisances?	No, based on extensive U.S. use with high explosives munitions.	No. Implosion in a vacuum in double-walled chamber reduces noise and vibration. No odors.	No.	No.
Would the process be able to satisfy environmental regulatory requirements under RCRA?	Likely. Has been approved by regulators at least once under RCRA for destruction of conventional munitions.	Likely.	Likely.	Yes. The EDS has already received regulatory approval in several different states.
Would the process be able to satisfy environmental regulatory requirements under the Clean Air Act?	Likely.	Likely.	Likely.	Yes. The EDS has already received regulatory approval in several different states.
Would the process be able to satisfy other applicable environmental regulatory requirements?	N/A. No other environmental regulatory requirements apply.	N/A. No other environmental regulatory requirements apply.	N/A. No other environmental regulatory requirements apply.	N/A. No other environmental regulatory requirements apply.

TABLE B-4 Continued

Public and Regulatory Acceptability in a U.S. Context Subfactors	Tier 1 International Munitions Processing Technologies CDC	DAVINCH	Dynasafe	U.S. EDS
Does the process satisfy the principles of pollution prevention and waste minimization?	On the basis of the information provided by the vendor, scrubbed gases and solids are the only wastes produced other than arsenic.	On the basis of the information provided by the vendor, scrubbed gases and metal are the only wastes produced other than arsenic.	Yes. Wastes are metal, scrubbed offgases, arsenic residues, and contaminated filter media.	The process satisfies the principles of pollution prevention. Although the process results in a large amount, relatively speaking, of secondary wastes, the amount is not excessive.
Overall	Has been used in U.S. for conventional munitions but not for CWM.	Ability to hold, analyze, and, if necessary, retreat offgases before release should increase acceptability to public and regulators.	Too early to tell since offgas treatment system may be viewed as being incineration-like.	Good record of acceptance.

TABLE B-5 Secondary Waste Issues Subfactor Evaluations for Tier 1 Munitions Processing Technologies

	Tier 1 International Munitions Processing Technologies			
Secondary Waste Issues Subfactors	CDC	DAVINCH	Dynasafe	U.S. EDS
What is the character of secondary wastes? Form (e.g., liquid, solid, gas) Volume Toxicity (e.g., agent, degradation products, metals, other contaminants)	Form: solids and gases. Solids include metal fragments, pea gravel and its dust, soot, spent lime, and filter carbon. The solids are to be decontaminated for release by hot air treatment.	Form: solids and gases. Gas volume is about 20,000 L/shot (volume of the inner vessel). Solids: 60 kg/shot for two Yellow bombs and 90 kg/shot if three Yellow bombs destroyed. Metal fragments claimed to meet GPL values for agents. Arsenic and arsenic oxides can remain on metal surfaces and on inner vessel walls.	Form: offgas from detonation and agent/energetic destruction. Solid metal scrap. Volume: gas volumes not known. Scrap volume depends on weight of munitions. Toxicity: metal claimed to be releasable as scrap. Offgas toxicity will vary with agent fill in munitions, but offgas can be extensively treated.	Solids: munition fragments. Liquids: neutralents and rinsates. Gases: carbon filter and vented after analysis.
Do secondary wastes initially meet: General population limits (GPLs) or short-term exposure limits (STELs)? CWC requirements? Environmental regulatory requirements?	Solids claimed to meet STELs. CWC requirements and environmental requirements were met in Belgium.	Offgases and metals claimed to meet GPLs for agents. CWC requirements and environmental requirements were met in Japan. To be determined in U.S.	Not known for secondary wastes. Scrap metal cleaned to meet GPL requirements. Neat chemical agents (mustard agent, Clark agents) have been destroyed in chamber, but these were not in chemical munitions. DRE of 99.9999 percent demonstrated. Ability to meet GPL, STEL, CWC, and environmental regulations not yet known.	Yes. Solid secondary wastes meet GPLs or STELS. With respect to CWC requirements, it is possible that neutralent could contain CWC Schedule 2 compounds above levels of concern to the CWC. Solid and gaseous secondary wastes meet CWC requirements. With respect to environmental requirements, liquids would likely require additional treatment to address hazardous waste characteristics of toxicity and corrosivity. Neutralent may also be deep-well injected. Gases and solids meet regulatory requirements.

TABLE B-5 Continued

Secondary Waste Issues Subfactors	Tier 1 International Munitions Processing Technologies: CDC	DAVINCH	Dynasafe	U.S. EDS
For each secondary waste, will subsequent treatment be required: To meet GPLs or STELs? To satisfy CWC requirements? To satisfy environmental regulatory requirements?	Yes, for offgases. No, for CWC. Solids may be recycled but may require additional treatment if contaminated with arsenic.	Yes, for offgases. Solids may be recycled but may require additional treatment if contaminated with arsenic.	Yes, for offgas. Solids may be recycled, but may require additional treatment if contaminated with arsenic.	Secondary waste treatment is unlikely to be necessary to reduce concentrations to GPLs or STELs for any of the EDS secondary wastes. With respect to CWC requirements, neutralent could contain CWC Schedule 2 compounds above levels of concern to the CWC, and in this case subsequent treatment would be required. Solid and gaseous secondary wastes meet CWC requirements. Solids, mainly munition fragments, would likely be able to be recycled, meeting regulatory requirements. Liquid wastes, if not otherwise disposed of (e.g., deep well injected), will require treatment at a TSDF to address the RCRA characteristics of reactivity and corrosivity.
For each secondary waste, if subsequent treatment is needed, are treatment methods established and available?	Yes. Well-established back end: gas scrubbing, catalytic oxidation, and carbon filtration. Solids are suitable for standard TSDF treatments or landfilling.	Yes. Well-established back end: gas scrubbing, combustion, and carbon filtration. Cold plasma is an alternative to offgas combustion, but track record is not known.	Yes for offgas standard cleanup process: cyclone, combustion chamber, carbon filtration used.	Yes, for both solids and liquids.

continued

TABLE B-5 Continued

	Tier 1 International Munitions Processing Technologies			
Secondary Waste Issues Subfactors	CDC	DAVINCH	Dynasafe	U.S. EDS
Will residuals from treatment of secondary waste require subsequent treatment: To meet GPLs or STELs? To satisfy CWC requirements? To satisfy environmental regulatory requirements?	Not likely unless arsenic is present in the waste residuals.	Not likely unless arsenic is present in the waste residuals.	Not likely unless arsenic is present in the waste residuals.	Residuals will not require additional treatment to address GPLs or STELs or to meet CWC requirements. Residuals from treatment of spent carbon may require regeneration or treatment to remove absorbed chemicals to satisfy environmental requirements. Also, residuals from treatment of neutralent may require further treatment, depending on the applied technology.
What is the disposition of final treatment residuals: Recycle? Hazardous waste landfill? Nonhazardous waste landfill? Other?	Metals can be recycled. Treated gases go to atmosphere. Other solids are likely to require disposal in a hazardous waste landfill. Arsenic may require final treatment and disposal.	Metals can be recycled. Treated gases released to atmosphere. Other solids are likely to require disposal in a hazardous waste landfill. Arsenic may require final treatment and disposal.	Metals can be recycled. Treated gases released to atmosphere. Other solids are likely to require disposal in a hazardous waste landfill. Fate of metallic arsenic and compounds not known.	Gaseous materials absorbed on spent carbon would be either regenerated or incinerated. Solids, primarily munition fragments, may be recycled or landfilled. If landfilled, the choice of hazardous or nonhazardous waste landfill would be dependent on state regulations. Final treatment for neutralent may include disposal of residual solids (e.g., incinerator ash) in a hazardous or nonhazardous waste landfill, depending on waste characteristics and state requirements. These liquids may also be treated using other technologies, such as solidification. Final treatment residuals may also be released under a Clean Water Act permit.
Overall	Some posttreatment of solid wastes may be needed before disposal.	No apparent secondary waste issues other than for arsenic cleanup and processing.	Standard offgas treatment required, and arsenic must be disposed of.	Wastes are handled by standard TSDF procedures.

Appendix C

Tier 1 Agent-Only Processing Evaluation Subfactor Comparative Tables

TABLE C-1 Process Maturity Subfactor Evaluations for Tier 1 Agent-Only Processing Technologies

	Tier 1 International Agent-Only Processing Technologies		
Maturity Subfactors	Russian Two-Stage Neutralization/ Bitumenization	Incineration	U.S.: RRS, SCANS, and Neutralization and Hydrolysis
Is the technology in use for any type of material, even one not related to CWM in the U.S.?	The neutralization/bitumenization technology has not been used in the U.S.	Incineration has been used and is being used extensively for both agent and secondary waste in the U.S. Due to access issues, however, incineration may not be the best choice for recovered munitions.	RRS and SCANS are used routinely in the U.S. Neutralization and hydrolysis have been used widely in the U.S.
Is the technology in use for any type of material, even non-CWM related internationally?	The technology is still in the developmental stage and has not been used internationally. Need to determine whether or not large-scale testing has been done in Russia.	Incineration has been used and is being used extensively for both agent and secondary waste in foreign countries.	RRS and SCANS have not been used internationally. Neutralization and hydrolysis have been widely used internationally.
Has the technology been permitted or otherwise approved in the U.S. for CWM or energetics?	The technology been not been permitted or otherwise approved in the U.S. for CWA or energetics.	Incineration has been permitted in the U.S. for the destruction of chemical weapons. In addition, commercial incinerators have been used to treat some types of secondary wastes.	The technology has been permitted or otherwise approved for use in the U.S.
Has the technology been permitted or otherwise approved in the U.S. for industrial wastes?	The technology has not been permitted or otherwise approved in the U.S. for industrial wastes.	Incineration has been permitted many times in the U.S. for the destruction of industrial wastes. It is a primary technology used to treat most organic industrial wastes in the U.S.	Neutralization and hydrolysis have been permitted or otherwise approved for use in the U.S. for industrial waste treatment.
How much, if any, additional R&D is required in order to implement the technology?	Additional R&D would be required to implement the technology in the U.S. Must locate or develop leachability data for bitumen-salt product.	No additional R&D would be required to implement the technology.	None
What, if any, are the scale-up requirements needed to implement the technology?	Depends on the scale of the testing at GOSNIIOCHT, which is not known.	No scale-up would be required.	None.
Can the technology be implemented within 3 to 5 years?	Depends on the scale of the testing at GOSNIIOCHT, which is not known, and the outcome of leachability testing.	The technology could be implemented within 3 to 5 years.	Yes.
Overall	More information is needed on the status of testing in Russia.	Incineration is a very mature technology but may not be applicable to recovered munitions.	In general, neutralization (including as used in the RRS and SCANS) and hydrolysis are very mature technologies.

TABLE C-2 Process Efficacy/Throughput Subfactor Evaluations for Tier 1 Agent-Only Processing Technologies

	Tier 1 International Agent-Only Processing Technologies		
Process Efficacy Subfactors	Russian Two-Stage Neutralization/ Bitumenization	Incineration	U.S.: RRS, SCANS, and Neutralization and Hydrolysis
What is the DRE?	Destruction efficiencies for the overall process are reported as greater than 99.999 percent. Residual agent concentrations in the final product are generally below the detection limit of 1×10^{-4} mg/ml.	The DRE for agent destruction is greater than 99.9999 percent.	RRS can treat to below detection limits of 25 ppb for GB, 50 ppm for mustard, 1 ppm for VX.
Does agent destruction meet the terms of the CWC (irreversible and verifiable)?	A joint Russian/U.S evaluation report states that the agent destruction meets the terms of the CWC (irreversible and verifiable) (see Chapter 5).	Agent destruction meets the terms of the CWC (irreversible and verifiable).	Agent destruction meets the terms of the CWC (irreversible and verifiable).
What is the DRE for energetics?	The technology does not address the destruction of energetics.	The DRE for energetics destruction is greater than 99.99 percent.	N/A
Is the process reliable and robust?	The process appears to be reliable and robust.	The technology is robust and can be used to destroy a variety of agents and energetics in a wide range of munitions.	The process is reliable and robust.
Is the process highly complex or relatively simple?	The process is very simple.	Complexity is variable, depending on application. As used for destruction of the U.S. stockpile, it is complex. As used in Iraq, it is not complex.	The process is very simple.
What are the personnel/staffing requirements for the technology?	The staffing requirements appear to be relatively low.	No unusual staffing requirements exist. Probably more than Russian two-stage process.	The staffing requirements are relatively low (27 for three-shift operations at Deseret Chemical Depot).
What is the process throughput?	The Shchuch'ye facility will have a capacity of 1,200 metric tons per year.	Process throughput can be very low (Canada, Belgium) or very high (Tooele).	Very low.
Is the process scalable so that it can address small, medium, and large munition finds?	The process appears to be scalable. Conventional liquid phase reactors are used, and these can be purchased in a wide range of sizes.	The process is scalable and can be used for essentially any size find. For very small finds, other technologies may be more appropriate. Alternatively, it can be combined with neutralization, or incineration can be used only to destroy neutralents (e.g., hydrolysates) or neutralents and energetics.	RRS and SCANS are intended for use on small quantities of CAIS. Neutralization and hydrolysis can be used for small or large finds. Neutralization and hydrolysis systems can be built at any desired capacity.
Is the process capable of handling multiple munition types?	The process treats agent only, not complete munitions.	The technology has been used to destroy multiple munition types. It has been used for non-stockpile-like materials internationally.	The process treats agent only, not complete munitions.
Is the process capable of handling multiple agent types?	The process has been developed for the Russian version of VX, called VR, for VX, for GB and GD, and for mustard agent.	The technology has been used to destroy multiple agent types.	The technology has been used to destroy multiple agent types.

continued

TABLE C-2 Continued

	Tier 1 International Agent-Only Processing Technologies		
Process Efficacy Subfactors	Russian Two-Stage Neutralization/ Bitumenization	Incineration	U.S.: RRS, SCANS, and Neutralization and Hydrolysis
Is the process transportable?	A transportable version has not been developed; however, there is apparently no reason this cannot be done.	The process can be either fixed or transportable, but in the U.S., transportable systems would likely face permitting obstacles.	The process can be either fixed or transportable.
Overall	For its intended purpose, destruction of nerve agents and mustard, the process appears to be effective.	The process can be used for a variety of applications and over a range of sizes.	For destruction of agent, neutralization and hydrolysis are very effective.

TABLE C-3 Process Safety Subfactor Evaluations for Tier 1 Agent-Only Processing Technologies

	Tier 1 International Agent-Only Processing Technologies		
Process Safety Subfactors	Russian Two-Stage Neutralization/ Bitumenization	Incineration	U.S.: RRS, SCANS, and Neutralization and Hydrolysis
What are the worker safety and health risks?	The process appears to offer no unusual safety hazards. Temperatures and pressures are moderate, etc.	Process is now very mature. Risks to workers, such as by exposure to agent or to accidents during maintenance, exist but are not excessive.	RRS, SCANS, and other neutralization and hydrolysis technologies offer no exceptional safety and health risks.
What are the community safety and health risks?	Community safety and health risks depend on precautions taken in implementing the technology.	Primarily exposure to agent. Minimal.	Minimal.
What are the process monitoring requirements?	Must be able to measure agent and Schedule 2 compounds in the distillate. Must be able to measure agent in the bitumen-salt product and in any offgases.	Extensive. The processes are complex. Monitoring for agent at emission points is critical.	Process monitoring is moderate.
To what extent have engineering controls been developed to ensure process safety?	Not known.	Engineering controls to ensure process safety have been extensively developed.	Engineering controls to ensure process safety have been extensively developed.
Overall	The process appears to be inherently safe.	With care, the process can be operated safely.	The process can be operated safely.

TABLE C-4 Public and Regulatory Acceptability Subfactor Evaluations for Tier 1 Agent-Only Processing Technologies

Public and Regulatory Acceptability in a U.S. Context Subfactors	Tier 1 International Agent-Only Processing Technologies: Russian Two-Stage Neutralization/ Bitumenization	Incineration	U.S.: RRS, SCANS, and Neutralization and Hydrolysis
Is the process inherently incineration-like?	The process is not inherently incineration-like.	The process is incineration.	The process is not inherently incineration-like.
Does the process break key chemical bonds (e.g., C-P bond for nerve agents)?	The process breaks P-S and P-F bonds, which cause the toxicity of nerve agents, but apparently not P-C bonds. However, Schedule 2 compounds with P-C bonds are immobilized in the bitumen-salt product.	The process breaks all key chemical bonds.	Not necessarily. Treatment of neutralents and hydrolysates may be needed for this purpose.
Could the process produce dioxins or other notable by-products?	The process does not produce dioxins. Other compounds present in the final bitumen-salt product might be of concern.	Dioxins and furans can be produced.	The process does not produce dioxins or furans.
Does the process allow holding and testing of process residuals prior to release?	The process allows holding and testing process residuals prior to release.	Hold-and-test is not used for the offgases, but liquid and solid secondary wastes can be held and tested prior to release for further management.	The process allows holding and testing process residuals prior to release.
Does the process result in excessive noise, odors, or other nuisances?	The odor of the bitumen-salt product needs to be checked. Otherwise, the process does not appear to result in excessive noise, odors, or other nuisances.	No.	No.
Would the process be able to satisfy environmental regulatory requirements under the RCRA?	Yes, very possibly. Leaching tests on the bitumen-salt product are needed.	Yes.	Yes.
Would the process be able to satisfy environmental regulatory requirements under the Clean Air Act (CAA)?	With proper engineering design for air pollution controls, it is likely that the process would be able to satisfy environmental regulatory requirements under the CAA.	Yes.	Yes.
Would the process be able to satisfy other applicable environmental regulatory requirements?	It does not appear that other environmental regulatory requirements would apply.	As evidenced by the obtaining of permits for the four mainland stockpile sites, the process is able to satisfy other applicable environmental regulatory requirements.	It does not appear that other environmental regulatory requirements would apply.
Does the process satisfy the principles of pollution prevention and waste minimization?	A large volume of secondary waste is produced that might or might not contain methylphosphonates and is landfilled, requiring continuing monitoring over time.	Secondary wastes produced as a result of incineration are generally low in volume as compared to the original materials. While emissions are generally low, public stakeholders have shown concern about the potential for some types of contaminants (e.g., metals, dioxins) to be released from the stack.	Relatively large volumes of secondary waste are produced.
Overall	Unknown for U.S. applications.	Public and regulatory acceptability is not always good.	Public and regulatory acceptability is generally good.

TABLE C-5 Secondary Waste Issues Subfactor Evaluations for Tier 1 Agent-Only Processing Technologies

	Tier 1 International Agent-Only Processing Technologies		
Secondary Waste Issues Subfactors	Russian Two-Stage Neutralization/ Bitumenization	Incineration	U.S.: RRS, SCANS, and Neutralization and Hydrolysis
What is the character of secondary wastes? Form (e.g., liquid, solid, gas) Volume Toxicity (e.g., agent, degradation products, metals, other contaminants)	The bitumen-salt final product is a solid and is produced at a volume several times that of the starting agent. More information is needed on the properties of that product, especially regarding leachability of degradation products, metals, and other contaminants.	Secondary waste comprise solids, liquids, and gas. Volumes of some streams, such as metal parts from the metal parts furnace, are large.	Neutralents and hydrolysates are liquids, usually of substantial volume and toxicity. Toxicity may, however, be due to the nature of the neutralizing (hydrolysis) material and not to chemical agent or degradation product content.
Do secondary wastes initially meet: General population limits (GPLs) or short-term exposure limits (STELs)? CWC requirements? Environmental regulatory requirements?	More information is needed on the properties of the bitumen-salt product.	While secondary wastes meet GPLs/STELs and CWC requirements, some secondary wastes require additional treatment to meet environmental regulatory requirements.	In general, neutralization treats agent to below detection capabilities. Treatment of neutralents and hydrolysates is typically needed to meet CWC and environmental regulatory requirements.
For each secondary waste, will subsequent treatment be required: To meet GPLs or STELs? To satisfy CWC requirements? To satisfy environmental regulatory requirements?	Unlikely. However, additional treatment to reduce leachability could be required.	While secondary wastes meet GPLs/STELs and CWC requirements, some secondary wastes require additional treatment to meet environmental regulatory requirements.	Treatment may be required to meet CWC and environmental regulatory requirements.
For each secondary waste, if subsequent treatment is needed, are treatment methods established and available?	Dependent on leachability concerns, if any.	Yes.	Treatment methods for neutralents and hydrolysates are established and available.
Will residuals from treatment of secondary waste require subsequent treatment: To meet GPLs or STELs? To satisfy CWC requirements? To satisfy environmental regulatory requirements?	Dependent on leachability concerns, if any.	Some final polishing steps may be needed for some secondary wastes, but such treatment is not uncommon for many industrial wastes.	Some final polishing steps may be needed for some secondary wastes, but such treatment is not uncommon for many industrial wastes.
How are the residuals of final treatment disposed of: Recycle? Hazardous waste landfill? Non-hazardous-waste landfill? Other?	In the U.S., the bitumen-salt final product would probably be sent to a hazardous waste landfill.	Final treatment residues are sent to various places, as shown in Table 5-2.	Various. Depends on treatment method (incineration, biotreatment, etc.).
Overall	Probably acceptable. More information is needed on the properties of the bitumen-salt product.	Management of secondary waste has become routine.	Management of secondary waste has become routine.

Appendix D

Committee Meetings and Other Activities

MEETINGS

First Committee Meeting: September 7-8, 2005, Bel Air, Maryland

Objective: Receive briefings from the Army, discuss and arrive at initial approach to task.

NSCMP Overview Briefing, William R. Brankowitz, Deputy Project Manager, Non-Stockpile Chemical Materiel Project

Explosive Destruction System (EDS) Lewisite and VX Testing, Trish Weiss, EDS Systems Manager, PMNSCMP

Non-Stockpile Mobile Technologies, Alan Seitzinger, Systems Manager, PMNSCMP

Non-Stockpile Neutralization Chemistry, Lucy Forrest, Chemistry Task Manager, PMNSCMP

Acid Digestion of GTRs Results and Path Forward, Edward Doyle, Alternative Systems Demonstration and Evaluation Group Leader, PMNSCMP

Second Committee Meeting: October 18-19, 2005, Washington, D.C.

Objective: Discuss evaluation factors, committee activities, report development, and future activities.

No briefings.

Third Committee Meeting: November 29-30, 2005, Washington, D.C.

Objective: Obtain information from the Army; discuss committee activities, report development, and future activities.

No briefings. An informal discussion was held with PMNSCM staff on a variety of topics pertinent to the committee's task.

Fourth Committee Meeting: January 31-February 1, 2006, Washington, D.C.

Objective: Receive data-gathering activity updates, discuss and develop report leading to a First Full Message Draft.

No briefings.

Fifth Committee Meeting: March 29-30, 2006, Washington, D.C.

Objectives: Review changes to the report since the March 2, 2006, preconcurrence draft, discuss and finalize report text, and attain committee concurrence.

No briefings.

DATA-GATHERING ACTIVITIES

Arlington, Virginia, September 22, 2005

Objective: Meet with representatives of the Environmental Security Technology Certification Program and the Strategic Environmental Research and Development Program to discuss nonintrusive, in-ground assessment technologies for possible use with large CWM burial sites.

Individuals met with: Jeff Marquesee, Environmental Security Technology Certification Program and Strategic Environmental Research and Development Program, and Anne Andrews, Environmental Security Technology Certification Program and Strategic Environmental Research and Development Program.

NRC participants: James Pastorick and Leonard Siegel, committee members.

Washington, D.C., November 11, 2005

Objective: Meet with representatives of Kobe Steel, Ltd., and GEOMET Technologies, LLC, to obtain information about Kobe Steel's DAVINCH controlled detonation munitions demilitarization process.

Individuals met with: Joseph Asahina and Masato Katayama, Kobe Steel, Ltd.; Tsuyoshi Imakita, Kobelco Research Institute, Inc.; and Frank Augustine, GEOMET Technologies, LLC.

NRC participants: Martin Gollin, Douglas Medville, and George Parshall, committee members; Harrison Pannella, study director; LaTanya Clemencia, project assistant; and James Myska, research associate.

Washington, D.C., November 18, 2005

Objective: Follow-up meeting with Kobe Steel representatives to obtain clarifications to questions raised by the November 11, 2005, meeting regarding the DAVINCH controlled detonation munitions demilitarization process.

Individual met with: Joseph Asahina, Kobe Steel, Ltd.

NRC participant: Douglas Medville, committee member.

Ashburn, Virginia, November 22, 2005

Objective: Meet with UXB International, Inc., to learn more about the Dynasafe Static Destruction Chamber munitions demilitarization process.

Individual met with: Harley Heaton, UXB International, Inc.

NRC participants: Martin Gollin and Douglas Medville, committee members, and James Myska, research associate.

Teleconference, December 8, 2005

Objective: Discuss international ACW demilitarization technologies, approaches, and issues with a representative of the Organisation for the Prohibition of Chemical Weapons (OPCW).

Individual met with: Jeff Osborne, OPCW.

NRC participants: Martin Gollin, Todd Kimmel, Douglas Medville, George Parshall, and James Pastorick, committee members; Harrison Pannella, study director; and James Myska, research associate.

Gaithersburg, Maryland, December 22, 2005

Objective: Discuss potential for DAVINCH implementation in the United States and the relevant permitting issues.

Individuals met with: Ted Prociv and Frank Augustine, GEOMET Technologies, LLC.

NRC participants: Todd Kimmel and Douglas Medville, committee members.

Porton Down, England, January 13, 2006

Objective: To ascertain the status of technologies used by the U.K. for the destruction of non-stockpile munitions, with a focus on incineration and the CDC, and to discuss a possible photocatalytic agent destruction technology.

Individuals met with: Richard Soilleux, Robert Cox, Nick Stokes, and Nigel Tonkin, Defence Science and Technology Laboratory.

NRC participant: Richard Ayen, committee chair.

Munster, Germany, January 16, 2006

Objective: Meet with representatives of Dynasafe to obtain further technical information on the Dynasafe static detonation chamber demilitarization technology.

Individuals met with: Rich Dugger IV and Harley Heaton, UXB; Johnny Ohlson, Dynasafe; Holger Weigel and Thomas Stock, Dynasafe Germany.

NRC participants: Douglas Medville and Martin Gollin, committee members.

Munster, Germany, January 16–17, 2006

Objective: Meet with representatives of the GEKA Munster chemical munitions demilitarization facility.

Individual met with: Hans-Joachim Grimsel, GEKA.

NRC participants: Douglas Medville and Martin Gollin, committee members.

The Hague, The Netherlands, January 18, 2006

Objective: Meet with representatives of the OPCW and obtain their opinions of various international chemical munition demilitarization technologies.

Individuals met with: Ralf Trapp, Jeff Osborne, and Jerzy Mazur, OPCW.

NRC participants: Douglas Medville and Martin Gollin, committee members.

Brussels, Belgium, January 19, 2006

Objective: Meet with representatives of the Belgian Royal Military Academy to obtain information on Belgian chemical munitions demilitarization work.

Individuals met with: Herbert DeBischopp and Michel Lefebvre, Belgian Royal Military Academy.

NRC participants: Douglas Medville and Martin Gollin, committee members.

Washington, D.C., January 30, 2006

Objective: Obtain technical information on the Controlled Detonation Chamber.

Individuals met with: Brint Bixler and John Coffey, CH2MHILL.

NRC participants: Richard Ayen, Martin Gollin, Todd Kimmel, George Parshall, and Jim Pastorick, committee members; Harrison Pannella, study director; James Myska, research associate.

Washington, D.C., February 3, 2006

Objective: Receive briefing on cold plasma, ask follow-up questions about DAVINCH capabilities, discuss munitions processing throughput, learn about evolution of DAVINCH since last meetings in November 2005.

Individuals met with: Joseph Asahina, Yasuhiro Morimoto, and Ryusuke Kitamura, Kobe Steel, Ltd.

NRC participants: Douglas Medville and James Pastorick, committee members; Harrison Pannella, study director; LaTanya Clemencia, project assistant.

Appendix E

Biographical Sketches of Committee Members

Richard J. Ayen (Chair)
Waste Management, Inc. (retired)

Dr. Ayen, now retired, was director of technology for Waste Management, Inc. He managed all aspects of Waste Management's Clemson Technical Center, including treatability studies and technology demonstrations for the treatment of hazardous and radioactive waste. His experience includes 20 years at Stauffer Chemical Company, where he was manager of the Process Development Department at Stauffer's Eastern Research Center. Dr. Ayen has published extensively in his fields of interest. He has significant experience in the evaluation and development of new technologies for the treatment of hazardous, radioactive, industrial, and municipal waste. Dr. Ayen was a member of the NRC Committee on Review and Evaluation of Alternative Technologies for Demilitarization of Assembled Chemical Weapons (I and II). He received his Ph.D. in chemical engineering from the University of Illinois.

Robin L. Autenrieth
Professor of Civil and Environmental Engineering
Texas A&M University

Dr. Autenrieth is a professor of civil and environmental engineering at Texas A&M University, received her B.S. degree in biological sciences from the University of Maryland, her M.S. degree in civil and environmental engineering from Clarkson College of Technology, and her Ph.D. in civil and environmental engineering from Clarkson University. She has a joint appointment in the TAMU Health Science Center's School of Rural Public Health. Dr. Autenrieth conducts research that connects engineering principles to the biological responses of environments exposed to damaging chemicals. Microbial biodegradation is one alternative to traditional remediation methods that rely on physically removing the contaminants or treating them on-site with neutralizing chemicals. Dr. Autenrieth's research on biodegradation kinetics on nerve and blister agents, as well as explosives and petroleum products, is being used to develop models to predict risks associated with exposure. She links environmental contamination to impact on exposed populations through human health risk assessments methods to estimate the potential for an adverse health effect.

Adrienne T. Cooper
Assistant Professor, Department of Civil and Environmental Engineering
Temple University

Dr. Cooper is an assistant professor in the Department of Civil and Environmental Engineering at Temple University. She has 20 years of experience in chemical and environmental engineering, including process engineering, process and waste treatment development, and environmental regulation. Dr. Cooper conducts research in catalytic processes for environmental treatment and remediation and pollution prevention. She is a recipient of the National Science Foundation's Early CAREER Award for her research on the development of photochemical reactors for water treatment and remediation. Dr. Cooper has served as a member of several non-stockpile technology evaluation panels since 1999. She holds a Ph.D. in environmental engineering from the University of Florida and a B.S. in chemical engineering from the University of Tennessee.

Martin Gollin
Process Design Engineering
Carmagen Engineering

Mr. Gollin is a process design and process safety consultant engineer with Carmagen Engineering, Inc., and was previously with ARCO Chemical Co. He has over 20 years of experience in process engineering and management of capital projects, risk assessment, process safety, loss prevention, and product development. From 1988 to 1999 he served as process design manager and principal engineer at ARCO Chemical Co., where he developed numerous processes and improvements. He was the EH&S manager for a $1 billion grass-roots project in the Netherlands and was a member of the panel that wrote the CCPS book *LOPA–Layer of Protection Analysis*. He earned B.S. and M.S. degrees in chemical engineering from Loughborough University of Technology in England.

Gary S. Groenewold
Staff Scientist
Idaho National Laboratory

Dr. Groenewold is a senior scientist who has conducted research in surface chemistry, gas-phase chemistry, and secondary ion mass spectrometry at the Idaho National Laboratory (INL) since 1991. His research has focused on determining the speciation of adsorbed radioactive and toxic metals (e.g., U, Pu, Hg, and Al) and organic compounds (e.g., VX, G agents, HD, organophosphates, amines, and sulfides). Prior to this, Dr. Groenewold served 3 years in line management at the INL and as the technical leader of an environmental organic analysis group. Before the INL, Dr. Groenewold worked in anticancer drug discovery for Bristol-Myers, using mass spectrometry as an identification tool. He received his Ph.D. in chemistry at the University of Nebraska, where he studied ion-molecule condensation and elimination reactions in the gas phase. He has authored 80 scientific publications on these subjects.

BG Paul F. Kavanaugh (ret.)
Engineering Management Consultant

General Kavanaugh, professional engineer, is an engineering management consultant. Previously, he was the director of government programs for Rust International, Inc., and director of strategic planning for Waste Management Environmental Services. In the Army, he served with the Army Corps of Engineers, the Department of Energy, the Defense Nuclear Agency, and managed facility upgrade projects at the U.S. Army Chemical Demilitarization Program at Johnston Atoll. He earned a B.S. in civil engineering from Norwich University and an M.S. in civil engineering from Oklahoma State University. He has expertise in military and civil works design and construction.

Todd A. Kimmell
Principal Investigator, Environmental Assessment Division
Argonne National Laboratory

Mr. Kimmell is principal investigator with the Environmental Assessment Division at the U.S. Department of Energy's Argonne National Laboratory. He is an environmental scientist and policy analyst, with more than 25 years' experience in solid and hazardous waste management, permitting and regulatory compliance, cleanup programs, and environmental programs and policy development. He has supported the Army's chemical weapons storage programs and has contributed to its Assembled Chemical Weapons Assessment Program and the Chemical Stockpile Emergency Preparedness Program. Mr. Kimmell also has a strong technical background in analytical and physical/chemical test method development and analytical quality assurance and control. He presently serves the Environmental Protection Agency's National Homeland Security Research Center on environmental test methods for chemical, biological, and radiological assessment for emergency response. Mr. Kimmell has also supported a number of environmental permitting programs at Army chemical weapons storage sites and at open burning/open denotation sites. He graduated from George Washington University with an M.S. in environmental science.

Loren D. Koller
Consultant
Environmental Health and Toxicology

Dr. Koller is an independent consultant and former professor and dean of the College of Veterinary Medicine at Oregon State University. His areas of expertise include pathology, toxicology, immunotoxicology, carcinogenesis, and risk assessment. He is a former member of the NRC Committee on Toxicology and has participated on several of its subcommittees, primarily involved in risk assessment. He served on the IOM Committee on the Assessment of Wartime Exposure to Herbicides in Vietnam and was invited to serve on committees for the CDC, EPA, Agency for Toxic Substances and Disease Registry (ATSDR), and the U.S. Army. He received his D.V.M. from Washington State University and his Ph.D. in pathology from the University of Wisconsin.

Douglas M. Medville
Program Leader
MITRE (retired)

Mr. Medville retired from MITRE as program leader for chemical materiel disposal and remediation. He has led many analyses of risk, process engineering, transportation, and alternative disposal technologies and has briefed the public and senior military officials on the results. Mr. Medville was responsible for evaluating the reliability and performance of the demilitarization machines used by the Army to disassemble stockpile chemical munitions and wrote several test plans and protocols for alternative chemical munition disposal technologies. He also led the evaluation of the operational performance of the Army's chemical weapon disposal facility on Johnson Atoll and directed an assessment of the risks, public perceptions, environmental aspects, and logistics of transporting recovered non-stockpile chemical warfare materiel to candidate storage and disposal destinations. Before that, he worked at Franklin Institute Research Laboratories and General Electric. Mr. Medville earned a B.S. in industrial engineering and an M.S. in operations research, both from New York University.

George W. Parshall (NAS)
Consultant
E.I. DuPont de Nemours & Company (retired)

Dr. Parshall retired from E.I. DuPont de Nemours & Company in 1992 after a career at the company spanning nearly 40 years. From 1979, he served as director of chemical science in Central Research and Development. Dr. Parshall is a past member of the NRC Board on Chemical Science and Technology and took part in earlier NRC chemical demilitarization studies. He continues to play an active role at the National Research Council. He graduated from the University of Illinois with a Ph.D. in organic chemistry.

James P. Pastorick
President
Geophex UXO, Ltd.

Mr. Pastorick is president of UXO PRO, Inc., an unexploded ordnance (UXO) consulting firm based in Alexandria, Virginia, that specializes in UXO planning and management consulting to state and foreign governments. Since he retired from the U.S. Navy as an explosives ordnance disposal officer and diver in 1989, he has been working on civilian UXO clearance projects. Prior to starting his present company, he was the senior project manager for UXO projects at UXB International, Inc., and the IT Group. He is a master rated unexploded ordnance technician with over 19 years of experience in explosive ordnance disposal.

Leonard M. Siegel
Director
Center for Public Environmental Oversight

Mr. Siegel is director of the Center for Public Environmental Oversight (CPEO) in Mountain View, California, a project of the Pacific Studies Center that facilitates public participation in the oversight of military environmental programs, federal facilities cleanup, and brownfields revitalization. One of the environmental movement's leading experts on military facility contamination, he serves on numerous advisory committees in that area, including the Interstate Technology Regulatory Council's work teams on vapor intrusion and perchlorate, the Moffett Field (formerly the Moffett Naval Air Station) Restoration Advisory Board, and the Outreach Advisory Committee of the Western Region Hazardous Substance Research Center. Mr. Siegel moderates and writes regularly for CPEO's Military Environmental Forum listserve.

William J. Walsh
Attorney and Partner
Pepper Hamilton LLP

Mr. Walsh is an attorney and partner in the Washington, D.C., office of Pepper Hamilton LLP. Prior to joining Pepper, he was section chief in the EPA Office of Enforcement. His legal experience encompasses environmental advice and environmental injury litigation involving a broad spectrum of issues pursuant to a variety of environmental statutes, including the Resources Conservation and Recovery Act (RCRA) and the Toxic Substances Control Act (TSCA). He represents trade associations, including the Rubber Manufacturers Association, in rule-making and other public policy advocacy; represents individual companies in environmental actions (particularly in negotiating cost-effective remedies in pollution cases involving water, air, and hazardous waste); and advises technology developers and users on taking advantage of the incentives for, and eliminating the regulatory barriers to, the use of innovative environmental technologies. He previously served on NRC committees concerned with the Non-Stockpile Chemical Agent Disposal Program, Superfund, and RCRA corrective action programs and the use of appropriate scientific groundwater models in environmental regulatory programs and related activities. Mr. Walsh holds a J.D. from George Washington University Law School and a B.S. in physics from Manhattan College.